热分析应用手册
Application Handbook
Thermal Analysis

药物和食品
Pharmaceuticals & Foods

Jürgen de Buhr　Georg Widmann　著
陆立明　译

本应用手册提供精选的应用实例,由瑞士梅特勒－托利多热分析实验室采用相关的仪器进行实验,作者根据最新知识进行数据处理。

但这并不意味着读者无需用适合试样的方法、仪器和用途进行亲自测试。由于对实例的效仿和应用是无法控制的,所以我们无法承担任何责任。

使用化学品、溶剂和气体时,必须遵循常规安全规范和制造商或供应商提供的使用指南。

This application handbook presents selected application examples. The experiments were conducted with the utmost care using the instruments specified in the description of each application at METTLER TOLEDO Themal Analysis Lab in Switzerland. The results have been evaluated according to the current state of our knowledge.

This does not however absolve you from personally testing the suitability of the examples for your own methods, instruments and purposes. Since the transfer and use of an application is beyond our control, we cannot of course accept any responsibility.

When chemicals, solvents and gases are used, general safety rules and the instructions given by the manufacturer or supplier must be observed.

图书在版编目(CIP)数据

药物和食品/(瑞士)德布尔,(瑞士)威德曼著;
陆立明译. —上海:东华大学出版社,2011.10
ISBN 978-7-81111-929-9

Ⅰ.①药… Ⅱ.①德… ②威… ③陆… Ⅲ.①热分析—手册 ②药物分析:热分析—手册 ③食品分析:热分析—手册 Ⅳ.①O657.7-62 ②TQ460.7-62 ③TS207.3-62

中国版本图书馆 CIP 数据核字(2011)第 177070 号

责任编辑:竺海娟
封面设计:杨 军

药物和食品

东华大学出版社出版
上海市延安西路1882号
邮政编码:200051 电话:(021)62373056
新华书店上海发行所发行 常熟大宏印刷有限公司印刷
开本:889×1194 1/16 印张:9.25 字数:296千字
2011年10月第1版 2019年12月第2次印刷
ISBN 978-7-81111-929-9
定价:68.00元

序

热分析是仪器分析的一个重要分支,它对物质的表征发挥着不可替代的作用。热分析历经百年的悠悠岁月,从矿物、金属的热分析兴起,近几十年在高分子科学和药物分析等方面唤起了勃勃生机。

我国在 20 世纪 50～60 年代,科研单位、高校和产业部门为满足科研、教学和生产的需要,经历了从原理出发自行设计研制热分析仪器的艰苦创业阶段;30～40 年前,先进的热分析仪器还只是在少数科研单位的测试中心才拥有,而随着我国综合国力的增强和对科研支持力度的加大,现已逐渐成为许多实验室的通用仪器,在科研和生产中起着愈来愈重要的作用。广大相关专业的科技人员为了更好地利用这些设备,迫切需要深入掌握热分析仪器及相关的基础和应用方面的知识。这套热分析应用系列丛书就是在这样的形势下应蕴而生的。

本书的基础数据主要是由瑞士的梅特勒－托利多(Mettler－Toledo)公司提供的,该公司是全球著名的精密仪器制造和经销商。早在 1945 年,就曾以首台单秤盘替代法天平而闻名于世。随后,又将其与加热炉结合,在 1964 年推出了世界上第一台商品化 TGA/DTA 热分析仪器。1968 年又有 TGA/MS 联用仪和差示扫描量热仪(DSC)相继问世。40 余年来,梅特勒－托利多一直是全球热分析仪器的主要供应商之一。现今具有包括 DSC、TGA/DSC、TMA、DMA 等完备的现代热分析仪器。近年取得的新进展有如:多星型热电堆 DSC 2006 年荣获美国 R&D100 奖,该奖项是每年颁发给当年在全球技术领域具有代表性新产品的开发者;2005 年开发的随机多频温度调制 DSC 技术 TOPEM™,能在一次实验中测定准稳态比热容,由热流与升温速率的相关性分析分别得到可逆、不可逆热流量和总热流量,以及反应(转变)过程与频率的关系。该公司最新推出的闪速 DSC(Flash DSC)更是热分析仪器的一项创举,为人们观测物质在快速升降温时的变化提供了新视解。

《热分析应用手册》系统介绍了热分析在诸多领域的应用。这套丛书是汇集了梅特勒－托利多公司瑞士总部和梅特勒－托利多(中国)公司科技人员的智慧而潜心编著的。《药物与食品》是其中的一个分册,热分析在药物熔点、纯度测定和多晶型观测等方面有极其广泛的应用。食品安全备受人们关注,按药食同源说,许多食品具有药物作用,含有某些生物活性成分,热分析可测定食品的组成、推测食品的新鲜程度和保持期,热分析是十分有效的快速检测食品的手段。在上述方面本书均给出了典型的应用示例。

该分册的主要作者 Jürgen de Buhr 和 Georg Widmann 长期在梅特勒－托利多瑞士热分析实验室工作,是热分析技术应用方面的资深专家。

译者陆立明先生 1985 年在华东理工大学获得聚合物材料工学硕士,后在上海市合成树脂研究所从事聚合物研究开发工作 12 年(其中 3 年在德国柏林技术大学进修高分子物理)。加入梅特勒－托利多(中国)公司以来一直从事热分析的技术应用和管理工作。

本书的酝酿和出版正值我国改革开放 30 年,我国在世界的影响全面提升。据报道这一时期我国科技人员在国际期刊发表的论文数量迅速提高。曾有报道,2004 年,中国科学院发表 SCI 论文较 1998 年增长 115%,总量已约为德国马普学会的 2 倍;在国际各领域居前 20 位学术刊物上发表的高质量论文数量,已占全国同期总量的一半以上(见:科学时报,2009 年 7 月 3 日第 2 版)。

本书的一个明显特点是以中英文对照的形式出版,这就为需要熟悉英语论文写作方法的读者提供了一种借鉴。

相信这套丛书的出版,将会对我国热分析技术的普及与提高起到重要的推动作用。

刘振海
2011 年 7 月于中国科学院长春应用化学研究所

出版前言

《热分析应用手册系列丛书》是由梅特勒－托利多瑞士热分析实验室专家撰写的系列手册，包括《热分析应用基础》、《热塑性聚合物》、《热固性树脂》、《弹性体》、《食品和药物》、《逸出气体分析》和《认证》等分册。

这套书既注重实用性，又注重学术性。可以将它们作为应用手册查询，也可以作为实验指南，如帮助选择合适的热分析测试技术和方法、制备和处理样品，设定实验参数等。手册中的所有应用实例都经过认真挑选，实验方法经过精心设计，测试曲线重复可靠，数据处理严格谨慎，结果解释和结论推导科学合理。

这套手册面向所有用到热分析和对热分析感兴趣的教授、科学家、工程师和学生（特别是研究生），适合所有热分析仪器的直接使用者。

本书是《热分析应用手册系列丛书》之《食品和药物》分册。英文原稿为《药物》和《食品》两个分册，我们将它们合并为一册出版。

热分析技术在药物以及食品行业具有广泛的应用，例如药物活性成分和非活性成分的分析表征、食物蛋白质变性的分析表征、安全性分析、稳定性研究以及质量控制中的日常分析。

本书简要介绍了 DSC（差示扫描量热法）、TGA（热重分析法）、TMA（热机械分析）和 DMA（动态热机械分析）等热分析的主要技术；通过许多实例，多方面深入介绍和讨论了热分析在药物和食品方面的应用。

与其他大多数分册一样，本书以中英文对照方式出版。本书无论对热分析工作者，还是对热分析学习者，应该都有帮助。

这里要特别感谢刘振海教授为本书作序。他仔细审阅了本书全部书稿，并逐字逐句进行修改，使本书的质量得到了很大提高。

同时感谢东华大学出版社编辑为出版本套丛书作出的辛勤努力。

译文甚至原著中，有错误之处，恳望读者指正，以便能在再版时改正，不胜感谢。

陆立明
2011 年 7 月，上海

目 录

1 热分析概论 Introduction to Thermal Analysis 1
  1.1 差示扫描量热法(DSC) Differential Scanning Calorimetry 1
    1.1.1 常规 DSC Conventional DSC 1
    1.1.2 温度调制 DSC(MTDSC) Temperature-modulated DSC 3
      1.1.2.1 ADSC 3
      1.1.2.2 IsoStep 4
      1.1.2.3 TOPEM™ 5
  1.2 热重分析(TGA) Thermogravimetric Analysis 6
  1.3 热机械分析(TMA) Thermomechanical analysis 8
  1.4 动态热机械分析(DMA) Dynamic Mechanical Analysis 9
  1.5 与 TGA 的同步测量 Simultaneous measurements with TGA 10
    1.5.1 同步 DSC 和差热分析(DTA、SDTA) Simultaneous DSC and Differential thermal analysis ... 10
    1.5.2 逸出气体分析(EGA) Evolved gas analysis 11
      1.5.2.1 TGA/MS 11
      1.5.2.2 TGA/FTIR 12

2 热分析在医药工业的应用 Applications of Thermal Analysis in the Pharmaceutical Industry 13
  2.1 热分析药物应用一览表 Application Overview Pharmaceuticals 13
  2.2 制药工业评说 Some Comments on the Pharmaceutical Industry 13
  2.3 热分析在药物上的应用 Applications of Thermal Analysis in Pharmaceuticals 14
    2.3.1 多晶型 Polymorphism 15
    2.3.2 假多晶型 Pseudopolymorphism 15
    2.3.3 相图 Phase diagrams 15
    2.3.4 稳定性 Stability 16
    2.3.5 相互作用 Interactions 16
    2.3.6 纯度测定 Purity determination 17
    2.3.7 包装材料 Packaging materials 17
    2.3.8 工艺优化 Process optimization 17
    2.3.9 校准和系统效应 Calibration, systematic effects 18
    2.3.10 一些重要概念和缩写 Some important Concepts, Abbreviations and Acronyms 18

3 热分析的药物典型应用 Typical TA Applications of Pharmaceuticals 21
  3.1 DSC 温度和热流量的校准 DSC Calibration, Temperature and Heat flow 21
  3.2 与升温速率无关的 DSC 校准 DSC Calibration, Hearting Rate Independence 22
  3.3 升温速率对丁基羟基茴香醚多晶型检测的影响 Influence of Heating Rate on the Detection of Polymorphism, Butylated Hydroxyanisole 24
  3.4 降温速率对蔗糖溶液结晶行为的影响 Influences of Cooling rate on Crystallization Behavior, Saccharose Solutions 25

I

3.5 升温速率对水包油乳膏水分含量测定的影响　Influences of Heating Rate on Moisture Content Determination, an O/W Cream ………………………………………………………………………… 27

3.6 升温速率对美托拉腙分解的影响　Influences of Heating Rate on Decomposition, Metolazone ………………………………………………………………………………………………… 28

3.7 坩埚对一水葡萄糖失水的影响　Influence of the Pan on Dehydration, Glucose Monohydrate ………………………………………………………………………………………………… 29

3.8 丁基羟基茴香醚的样品制备　Sample Preparation, Butylated Hydroxyanisole ……… 32

3.9 二丁基羟基甲苯试样量的影响　Influence of the Sample Weight, Butylated Hydroxytoluene ………………………………………………………………………………………… 33

3.10 样品贮存和吸湿效应　Sample Storage and Hygroscopic Effects ……………………… 34

3.11 油的氧化稳定性　Oxidation Stability of Oils ……………………………………………… 36

3.12 香草醛熔融行为的表征　Characterization of the Melting Behavior, Vanillin ………… 37

3.13 胆固醇十四烷酸酯的相转变　Phase Changes, Cholesteryl Myristate ………………… 39

3.14 根据熔融行为对聚乙烯醇的鉴别　Identification Based on Melting Behavior, Polyethylene Glycol ………………………………………………………………………………………… 40

3.15 糖溶液水的熔点降低　Melting Point Depression of Water, Sugar Solutions ………… 42

3.16 油包水乳膏的DSC"指纹"　DSC 'Fingerprint', O/W Cream …………………………… 44

3.17 D,L丙交酯-乙交酯共聚物的玻璃化转变　Glass Transition, Poly (D, L-lactide)-Co-Glycolide (DLPLGGLU) ……………………………………………………………………………… 45

3.18 羟丙基甲基纤维素邻苯二甲酸酯(HPMC-PH)的玻璃化转变和水分含量　Glass Transition and Moisture Content, Hydroxypropoxymethylcellulose Phthalate (HPMC-PH) ……… 46

3.19 聚乙烯薄膜的质量控制　Quality Control, PE Films ……………………………………… 49

3.20 氢化可的松的分解　Decomposition, Hydrocortisone ……………………………………… 50

3.21 甲磺酸双氢麦角胺熔点处的分解　Decomposition at the Melting Point, Dihydroergotamine Mesylate …………………………………………………………………………………………… 52

3.22 阿斯巴甜的熔融和分解　Melting Behavior and Decomposition, Aspartame ………… 53

3.23 丙二酸的完全分解　Total Decomposition, Malonic Acid ………………………………… 55

3.24 乙酰水杨酸分解的动力学分析　Kinetic Analysis of Decomposition, Acetylsalicylic Acid …… 57

3.25 茶碱的水合稳定性　Hydrate Stability, Theophylline …………………………………… 59

3.26 淀粉/羟甲基纤维素钠(羧甲基淀粉钠)的水分　Moisture, Starch/NaCMC (Primojel) …… 61

3.27 三棕榈精的多晶型　Polymorphism, Tripalmitin ………………………………………… 63

3.28 甲苯磺丁脲的多晶型　Polymorphism, Tolbutamide ……………………………………… 64

3.29 退火处理丁基羟基茴香醚多晶型　Polymorphic Modifications by Annealing, Butylated Hydroxyanisole ……………………………………………………………………………………… 66

3.30 硬脂酸镁的DSC"指纹"　DSC 'Fingerprint', Magnesium Stearate …………………… 67

3.31 左旋聚丙交酯的多晶型　Polymorphism, L-Polylactide ………………………………… 69

3.32 磺胺吡啶的多晶型　Polymorphism, Sulfapyridine ……………………………………… 70

3.33 一水葡萄糖的假多晶型　Pseudopolymorphism, Glucose Monohydrate ……………… 72

3.34 布洛芬(异丁苯丙酸)的光学纯度　Optical Purity, Ibuprofen ………………………… 74

3.35 对羟基苯甲酸及其酯的纯度测定（DSC法和HPLC法） Purity using DSC and HPLC, 4-Hydroxybenzoic Acid and its Esters ……………………………………………………… 76

3.36 非那西汀＋对氨基苯甲酸纯度测定 Purity Determination, Phenacetin ＋ 4-Aminobenzoic Acid ………………………………………………………………………………………… 78

3.37 胆甾醇的纯度和重结晶 Purity and Recrystallization, Cholesterol ……………………… 80

3.38 甲苯磺丁脲和聚乙二醇6000的相图 Phase Diagram, Tolbutamide and PEG 6000 …… 82

3.39 对羟基苯甲酸甲酯和对羟基苯甲酸的共熔体组成 Eutectic Composition, Methyl-4-Hydroxybenzoate and 4-Hydroxybenzoic Acid …………………………………………… 84

3.40 药物活性物质的TGA-MS溶剂检测 Solvent Detection by means of TGA-MS, Pharmaceutically Active Substance ………………………………………………………………………… 86

3.41 不同水分含量的油包水乳膏的定量分析 Quantification, O/W Creams with Different Water Content ……………………………………………………………………………… 88

3.42 一水茶碱的定量分析 Quantification, Theophylline Monohydrate ……………………… 90

3.43 Alcacyl中活性物质的测定 Determination of an Active Substance, Alcacyl …………… 92

4 热分析在食品工业的应用 Applications of Thermal Analysis in the Food Industry …… 95

4.1 热分析食品应用一览表 Application Overview Food ………………………………… 95

4.2 食品工业与热分析 Food Industry and Thermal Analysis …………………………… 95

4.2.1 食品工艺中的反应和相 Reactions and Phases in Food Technology ……………… 95

4.2.2 食品中主要成分DSC检测一览表 List of DSC Investigations of the Main Components in Foods …………………………………………………………………………… 96

4.2.3 蛋白质 Proteins ……………………………………………………………………… 97

4.2.4 碳水化合物 Carbohydrates ………………………………………………………… 99

4.2.5 脂肪和油 Fats and Oils …………………………………………………………… 100

4.2.6 食品包装材料——塑料薄膜 Food Packagings-Plastic Films …………………… 102

5 热分析食品的典型应用 Typical TA Applications of Food ………………………………… 104

5.1 植物蛋白质的变性 Denaturation of Vegetable Proteins ……………………………… 104

5.2 鸡蛋蛋白质的变性 Egg Protein Denaturation ………………………………………… 106

5.3 鸡蛋蛋清热处理的影响 Influence of Thermal Treatment of Egg White …………… 108

5.4 鸡蛋贮存时间的影响 Influence of Egg Storage Time ………………………………… 109

5.5 pH对牛血红蛋白的影响 Influence of pH on Bovine Hemoglobin ………………… 111

5.6 肉类的DSC DSC of Meat ……………………………………………………………… 113

5.7 淀粉的凝胶化 Gelatinization of Starch ………………………………………………… 114

5.8 水中淀粉含量对溶胀的影响 Influence of the Starch Content on Swelling in Water … 115

5.9 无定形糖的ADSC（调制DSC） ADSC of Amorphous Sugar ……………………… 117

5.10 糖和淀粉的TGA TGA of Sugar and Starch ………………………………………… 119

5.11 意大利通心粉的动态负载TMA Dynamic Load TMA of Pasta …………………… 121

5.12 巧克力的熔融 Melting of Chocolate ………………………………………………… 122

5.13 可可脂的热表征 Thermal Characterization of Cocoa Butter ……………………… 124

5.14 熔融行为和氢化作用 Melting Behavior and Hydrogenation ……………………… 128

5.15 植物油的结晶　Crystallization of Vegetable Oils ………………………………………… 129
5.16 棕榈油的液相含量和滴点　Liquid Fraction and Dropping Point of Palm Oils ………… 131
5.17 植物脂肪的氧化　Oxidation of Vegetable Fats ………………………………………… 133
5.18 乙醇/水混合物　Ethanol/Water Mixtures ……………………………………………… 134
5.19 塑料薄膜的鉴别　Identification of Plastic Films ……………………………………… 136

1 热分析概论 Introduction to Thermal Analysis

热分析是测试材料的物理和化学性能与温度关系的一类技术的总称。在所有这些方法中,样品受加热、冷却或等温温度程序控制。

测试可在不同气氛中进行,通常使用惰性气氛(氮气、氩气、氦气)或氧化气氛(空气、氧气)。有时在测试期间从一种气氛切换到另一种气氛。另一个可选择的参数是气体压力。

DSC 还可与能同步观察试样的仪器联用(DSC/显微镜法),或用不同波长的光照射(光量热法)。

Thermal analysis is the name given to a group of techniques used to measure the physical and chemical properties of materials as a function of temperature. In all these methods, the sample is subjected to a heating, cooling or isothermal temperature program.

The measurements can be performed in different atmospheres. Usually either an inert atmosphere (nitrogen, argon, helium) or an oxidative atmosphere (air, oxygen) is used. In some cases, the gases are switched from one atmosphere to another during the measurement. Another parameter sometimes selectively varied is the gas pressure.

DSC can also be used in combination with instruments that allow the sample to be simultaneously observed (DSC microscopy) or exposed to light of different wavelengths (photocalorimetry).

1.1 差示扫描量热法(DSC) Differential Scanning Calorimetry

DSC 测量流入和流出试样的热流量。DSC 可用于研究物理转变(玻璃化转变、结晶、熔融和挥发化合物的蒸发)和化学反应的热效应,所获得的信息可表征样品的热性能和组成。此外,还能测定诸如热容、玻璃化转变温度、熔融温度、反应热和反应程度等的性质。

In DSC, the heat flow to and from the sample is measured. DSC can be used to investigate thermal events such as physical transitions (the glass transition, crystallization, melting, and the vaporization of volatile compounds) and chemical reactions. The information obtained characterizes the sample with regard to its thermal behavior and composition. In addition, properties such as the heat capacity, glass transition temperature, melting temperature, heat and extent of reaction can also be determined.

1.1.1 常规 DSC Conventional DSC

常规 DSC 采用线性温度程序,试样和参比物(通常只是空坩埚)以线性速率升(降)温,或经常将几个局部程序即所谓的程序段连接在一起,形成一个完整的温度程序。典型的 DSC 曲线如图 1.1 所示。测试开始时曲线的变化是由于初始的"启动偏移"(曲线1)。在该瞬变区域,状态突然从等温模式变为线性升温模式。启动偏移的大小取决于试样热容和升温速率。试样中如果存在挥发性物质如溶剂,会观察到由于蒸发产生的吸热峰(曲线2),试样

Conventional DSC employs a linear temperature program. The sample and reference material (or just an empty crucible) are heated or cooled at a linear rate, or in some cases, held at a constant temperature (i. e. isothermally). Often several partial programs or so-called segments are joined together to form a complete temperature program. A typical DSC curve is shown schematically in Figure 1.1. The change in the curve at the beginning of the measurement is due to the initial "startup deflection" (1). In this transient region, the conditions suddenly change from an isothermal mode to a linear heating mode. The magnitude of the startup deflection depends on the heat capacity of the sample and the heating rate. If volatile substances such as solvents are present in the sample, an endothermic peak (2) is

失重。可通过称量测试前后的试样质量和使用不同种类的坩埚得到关于这种峰的更多信息。与开口坩埚不同,完全密封的坩埚可防止试样的蒸发。无热效应的 DSC 曲线部分(曲线 3),由于试样的热容通常线性增大,因而观察到称为"基线"的直线。熔融产生吸热峰(曲线 4)。最后,在较高的温度,开始分解(曲线 5)。实验所用吹扫气体的种类经常会影响发生的反应,尤其在高温下更是如此。

observed due to the vaporization; the sample loses mass. Further information on such peaks can be obtained by weighing the sample before and after the measurement and by using different types of crucibles. In contrast to open crucibles, hermetically sealed crucibles prevent vaporization of the sample. At the part of DSC curve with no thermal effects (3), the heat capacity of the sample increases normally linearly and therefore a straight line called "baseline" is observed. The melting produces endothermic peak (4). Finally, at higher temperatures, decomposition begins (5). The type of purge gas used in the experiment often has an influence on the reactions that occur, especially at high temperatures.

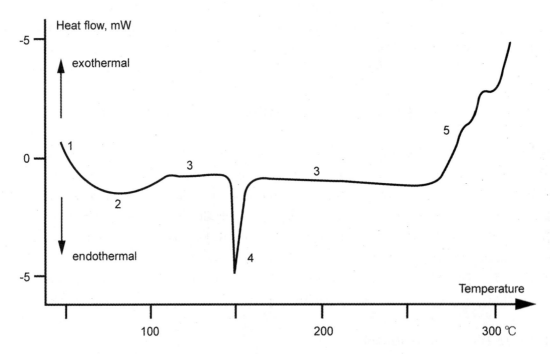

图 1.1　DSC 曲线示意图
1 初始启动偏移;2 水汽蒸发;3 基线(无热效应的 DSC 曲线部分);4 熔融峰;5 分解开始
Figure 1.1　Schematic DSC curve
1 initial startup deflection; 2 evaporation of moisture; 3 part of DSC curve with no thermal effects, i.e., baseline; 4 melting peak; 5 beginning of decomposition.

可通过冷却和再次测试试样来区分物理转变与化学反应—化学反应是不可逆的,而熔化了的结晶物质当冷却或二次升温时会重新结晶。玻璃化转变也是可逆的,但经常在玻璃化转变的第一次升温测试中观察到的焓松弛是不可逆的。

Physical transitions and chemical reactions can be differentiated by cooling the sample and measuring it again-chemical reactions are irreversible whereas crystalline materials melt then crystallize again on cooling or on heating a second time. Glass transitions are also reversible but not the enthalpy relaxation often observed in the first heating measurement of a glass transition.

1.1.2 温度调制 DSC(MTDSC)　Temperature-modulated DSC

1.1.2.1 ADSC

调制 DSC(ADSC)是一种专门的温度调制 DSC(MTDSC)。与常规 DSC 不同,在线性温度程序上叠加一个小的周期性温度变化。温度程序可以基础升温速率、温度振幅和周期性变化温度的持续时间来表征(图 1.2)。对于准等温测试,基础升温速率也可为零。

Alternating DSC (ADSC) is a particular type of temperature-modulated DSC (TMDSC). In contrast to conventional DSC, the linear temperature program is overlaid with a small periodic temperature change. The temperature program is characterized by the underlying heating rate, the temperature amplitude and the duration of the periodically changing temperature (Fig. 1.2). With quasi-isothermal measurements, the underlying heating rate can also be zero.

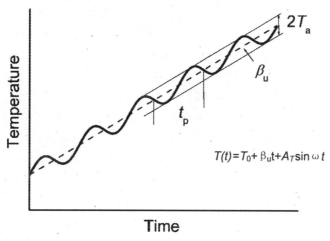

图 1.2 典型的 ADSC 温度程序

β_u 为基础升温速率,A_T 为温度振幅,t_p 为周期。$2\pi/P$ 为角频率 ω,P 为正弦波的周期

Figure 1.2　Typical ADSC temperature program

β_u is the underlying heating rate, A_T the temperature amplitude,

t_p period. The angular frequency ω is defined as $2\pi/P$ where P denotes the period of the sine wave.

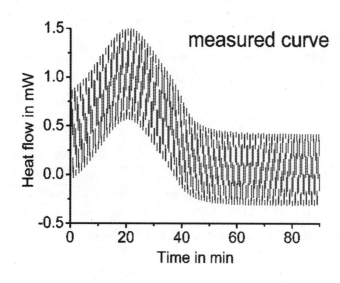

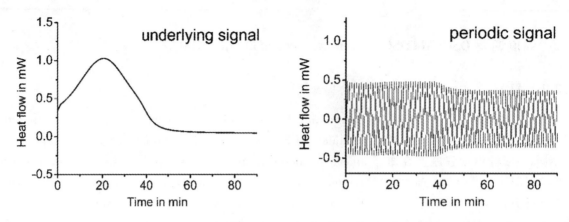

图 1.3 测得的 ADSC 曲线分离成基础和周期性信号成分
Figure 1.3　Separation of the measured ADSC curve into the underlying and the periodic signal components

温度调制的结果,是所测得的热流量周期性变化。该热流量可分成两部分,如图 1.3 所示。由信号平均得到基础信号(总热流),它相当于常规 DSC 曲线。作为附加信息,还得到周期性信号分量。可逆热流量为能够直接跟上升温速率变化的热流分量,从同相热容计算得到。总热流量减去可逆热流量得到不可逆热流量。该技术的优势之一是可将同时发生的过程分开。例如,可直接测量化学反应过程中的热容变化。

ADSC 曲线的计算以傅立叶分析为基础。复合热容 c_p^* 的复数模用下面的等式计算:

$$|c_p^*| = \frac{A_\Phi}{A_\beta} \cdot \frac{1}{m}$$

式中 A_Φ 和 A_β 为调制热流量和升温速率的振幅,m 为试样质量。ADSC 热流信号与升温速率之间的相角用于计算同相的 c_p。

1.1.2.2　IsoStep

IsoStep 是一种特殊的温度调制 DSC。这种方法温度程序是由很多开始和结束为等温段的动态程序段组成(图 1.4)。

As a result of temperature modulation, the measured heat flow changes periodically. This can be separated into two parts as shown in Figure 1.3. Signal averaging yields the underlying signal (total heat flow), which corresponds to the conventional DSC curve. As additional information, one also obtains the periodic signal component. The reversing heat flow corresponds to the heat flow component that is able to follow the heating rate change directly and is computed from the in-phase heat capacity. The difference between the total heat flow and the reversing heat flow yields the non-reversing heat flow. One advantage of this technique is that it allows processes that occur simultaneously to be separated. For example, the change in heat capacity during a chemical reaction can be measured directly.

The evaluation of the ADSC curves is based on Fourier analysis. The modulus of the complex heat capacity c_p^* is calculated using the equation

$$|c_p^*| = \frac{A_\Phi}{A_\beta} \cdot \frac{1}{m}$$

where A_Φ and A_β denote the amplitudes of the modulated heat flow and heating rate, and m the sample mass. The phase angle between the ADSC heat flow signal and the heating rate is used to calculate the in-phase c_p.

1.1.2.2　IsoStep

IsoStep is a special type of temperature-modulated DSC. In this method, the temperature program consists of a number of dynamic segments that begin and end with an isothermal segment (Fig. 1.4).

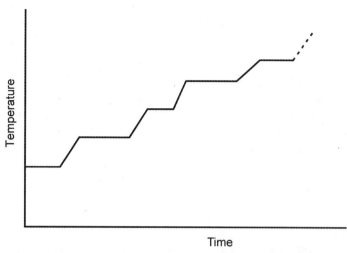

图 1.4 IsoStep 温度程序由不同的恒温和动态段组成

Fig. 1.4　IsoStep temperature program consisting of different isothermal and dynamic segments

等温段能让动态段的等温漂移获得修正，结果得到更好的热容准确性。等温步阶也可能包含动力学信息，例如化学反应。热容可用蓝宝石参比样进行测定，并可从热容变化区分开动力学效应。

The isothermal segments allow the isothermal drift of the dynamic segments to be corrected. This results in better heat capacity accuracy. The isothermal step may also contain kinetic information, for example of a chemical reaction. Heat capacity determinations can be made using a sapphire reference sample, and kinetic effects can be separated from changes in heat capacity.

1.1.2.3　TOPEM™

TOPEM™ 是高级温度调制 DSC 技术，基于 DSC（仪器和试样两者）对随机调制基础温度程序响应（图 1.5）的全面数学分析而设计的。由于温度脉冲是随机分布的，系统的温度振荡是在宽频范围，而不是只在某单一频率（例如 ADSC）。振荡式输入信号（升温速率）和响应信号（热流量）的相关分析能得到比常规温度调制 DSC 更多的信息，不仅能将可逆与不可逆效应分开，而且还能测量试样的准稳态热容和测定与频率有关的热容值。这能用来在一次单独的测试中就区分开频率有关的松弛效应（例如玻璃化转变）和与频率无关的效应（例如化学反应）。

TOPEM™ is an advanced temperature-modulated DSC technique that is based on the full mathematical analysis of the response of a DSC (both the apparatus and the sample) to a stochastically modulated underlying temperature program (Fig. 1.5). Due to the randomly distributed temperature pulses, the system is subjected to temperature oscillations over a wide frequency range and not just at one single frequency (e. g., ADSC). An analysis of the correlation of the oscillating input signal (heating rate) and the response signal (heat flow) provides much more information than can be obtained using conventional temperature-modulated DSC. Not only can reversing and non-reversing effects be separated, but the quasi-static heat capacity of the sample is also measured and frequency-dependent heat capacity values are determined. This can be used to distinguish between frequency-dependent relaxation effects (e. g. glass transitions) and frequency-independent effects (e. g. chemical reactions) in one single measurement.

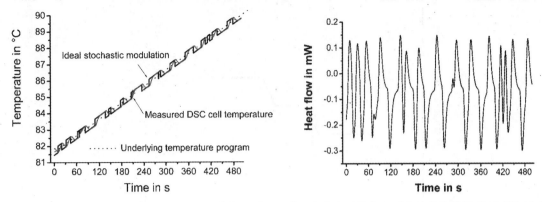

图 1.5 TOPEMTM方法炉体设定值的温度曲线(黑线)和升温速率在平均值上下变动的炉体温度(灰线)(左图)流向试样的热流量的不规则变动(右图)

Fig 1.5 Temperature curve of the furnace set value (black line) in a TOPEMTM method in which the furnace temperature (gray curve) generates a heating rate that fluctuates around a mean value. The heat flow to the sample also fluctuates irregularly as shown in the diagram on the right.

1.2 热重分析(TGA)　Thermogravimetric Analysis

当试样被加热时，经常开始失重。失重可能产生于蒸发，或形成逸出气体产物的化学反应。如果吹扫气氛是非惰性的，试样还可能与气体反应。在某些情况下，试样质量也可能增加，例如氧化反应，如果形成的产物是固体。

热重分析(TGA)是测量试样质量随温度或时间的变化。

由TGA可获得关于试样性质及其成分的信息。如果试样分解产生于化学反应，则试样质量通常呈台阶状变化。台阶出现时的温度可表征该样品在所用气氛中的稳定性。

典型的TGA曲线如图1.6所示。通过分析单独质量台阶的温度和高度可确定材料的组成。

水、残留溶剂等挥发性化合物在相对低的温度逸出(曲线1)。这些化合物的排除与气体压力有关，在低压下(真空)，相应的失重台阶向低温移动，蒸发加速。通常在比挥发性化合物蒸发温度更高的温度失去结晶水(曲线2)。

更高温度，则试样的主成分发生分解，形成一个较大的台阶(曲线3)。

When a sample is heated, it often begins to lose mass. This loss of mass can result from vaporization or from a chemical reaction in which gaseous products are formed and evolved from the sample. If the purge gas atmosphere is not inert, the sample can also react with the gas. In some cases, the sample mass may also increase, e.g. in an oxidation reaction if the product formed is a solid.

In thermogravimetric analysis (TGA), the change in mass of a sample is measured as a function of temperature or time.

TGA provides information on the properties of the sample and its composition. If the sample decomposes as a result of a chemical reaction, the mass of the sample often changes in a stepwise fashion. The temperature at which the step occurs characterizes the stability of the sample material in the atmosphere used.

Figure 1.6 shows a typical TGA curve. The composition of a material can be determined by analyzing the temperatures and the heights of the individual mass steps.

Volatile compounds such as water, residual solvents or added oils are evolved at relatively low temperatures. The elimination of such components depends on the gas pressure. At low pressures (vacuum), the corresponding mass loss step is shifted to lower temperatures, that is, vaporization is accelerated. ? Loss of water of crystallization occurs usually above the temperature at which volatile compounds evalporate (2).

At higher temperature, the main component decompose and produces a rather high step (3). The gases used can be air or

所用气氛可以是空气或氧气,也可以是氮气等惰性气体。分析在惰性气氛中的热解反应可从台阶高度确定组分含量,甚至可确定是何种物质。

由残留物可测定填料或灰分。由于浮力效应和气流速率而产生的测试曲线的微小变化,可通过减去空白曲线得以修正。

TGA 测试结果经常用 TGA 曲线的一阶微商(称为 DTG 曲线)表示。于是,TGA 曲线质量损失的台阶,则在 DTG 曲线呈现峰形。DTG 曲线相当于试样质量变化的速率。

oxygen, or inert gases such as nitrygen. The analysis of pyrolysis reactions in an inert atmosphere allows the content (from the step height) and possibly even the type of material to be determined.

The filler or ash is determined from the residue. Small changes in the measurement curve due to buoyancy effects and gas flow rate can be corrected by subtracting a blank curve.

TGA measurements are often displayed as the first derivative of the TGA curve, the so-called DTG curve. Steps due to loss of mass in the TGA curve then appear as peaks in the DTG curve. The DTG curve corresponds to the rate of change of sample mass.

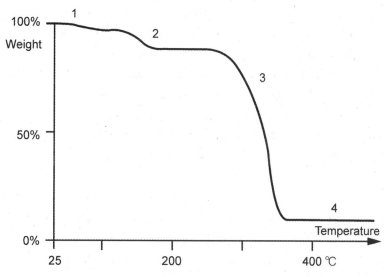

图 1.6　TGA 曲线示意图

1 挥发性成分蒸发的失重(水分、溶剂);2 失结晶水;3 分解;4 残留物(灰分、填料、在惰性气氛中生成的炭黑或烟灰)

Figure 1.6　Schematic TGA curve

1 loss of mass due to the vaporization of volatile components (moisture, solvents); 2 loss of water of crystallization; 3 decomposition; 4 residue (ash, fillers, carbon black or soot formed during decomposition in an inert atmosphere).

分解台阶的温度范围,在一定程度上受气体产物扩散性(即从试样扩散出去的容易程度)的影响。当使用反应性气体时,试样表面气体交换的效率是关键。可以使用合适的坩埚(例如 30 μL 氧化铝等浅皿坩埚)和合适的试样几何形状(几个小颗粒或粉末)来降低测试时的扩散效应。

TGA 非常精确地测量试样质量的变化。然而令人遗憾的是,该技术不提供有关逸出气体分解产物性质

The temperature range of the decomposition steps is influenced to a certain extent by the ease with which the gaseous products are able to diffuse out of the sample. When reactive atmospheres are used, the efficiency of gas exchange at the surface of the sample is crucial. The effects of diffusion on the measurement can be reduced by using suitable crucibles (e.g. crucibles with low wall-heights such as the 30-μL alumina crucible) and by suitable sample geometry (several small pieces or powder).

In TGA, the change in mass of the sample is measured very accurately. Unfortunately, however, the technique does not provide any information about the nature of the gaseous

的任何信息。不过,通过 TGA 与合适的气体分析仪耦联(逸出气体分析 EGA)可分析这些产物。

decomposition products evolved. The products can however be analyzed by coupling the TGA to a suitable gas analyzer (evolved gas analysis, EGA).

1.3 热机械分析(TMA)　Thermomechanical analysis

热机械分析测试试样升温时的尺寸变化。该项技术,连续测量带一定力放置于试样表面的探头的位置或位移与温度或时间之间的关系。图 1.7 所示为典型的 TMA 曲线。探头施加的压力和试样的硬度决定了 TMA 实验事实上是膨胀还是穿透测试。

Thermomechanical analysis measures the dimensional changes of a sample as it is heated. In this technique, the position or displacement of a probe resting on the surface of the sample with a certain force is continuously measured as a function of temperature or time. Figure 1.7 shows a typical TMA curve. The pressure exerted by the probe and the hardness of the sample determine whether the TMA experiment is in fact an expansion or a penetration measurement.

对于热膨胀测,探头在试样表面仅施加低压力。试样在所关心的温度范围内线性升温。直接由测试曲线计算线性热膨胀系数(CTE)。

In the thermal expansion measurement, the probe exerts only a low pressure on the surface of the sample. The sample is heated linearly over the temperature range of interest. The linear coefficient of thermal expansion (CTE) is calculated directly from the measurement curve.

针入实验探头施加的压力大得多。可直接测量试样升温时的软化温度。物质在玻璃化转变温度处或熔融时软化。

In a penetration experiment, the probe exerts a much greater pressure. The softening temperature can be directly measured when the sample is heated. Materials soften at the glass transition temperature or on melting.

如果对试样施加周期性变化的力,试样尺寸也周期性变化。该测试模式称为动态负载 TMA(DLTMA)。从振幅和试样厚度能估算出试样的弹性模量(杨氏模量)。

If a periodically changing force is applied to the sample, the sample dimensions also change periodically. This measurement mode is called dynamic load TMA, DLTMA. The elastic modulus (Young's modulus) of the sample can be estimated from the amplitude and the sample thickness.

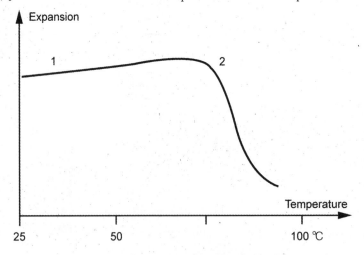

图 1.7　TMA 曲线示意图
1 低于玻璃化转变时的膨胀;2 软化(塑性形变)

Figure 1.7　Schematic TMA curve

1 expansion below the glass transition; 2 softening (plastic deformation).

1.4 动态热机械分析(DMA)　Dynamic Mechanical Analysis

动态热机械分析是测量力学模量与温度、频率和振幅的关系。

施加于试样的周期性(通常为正弦)变化的力在试样中产生周期性的应力。试样对该应力作出反应,仪器测量相应的形变行为。由应力和形变测定力学模量 M。取决于所加应力的类型,测定剪切模量 G(施加剪切应力)或杨氏模量 E(拉伸或弯曲)。

试样并不总是立即对周期性变化的应力作出反应——与样品的黏弹性有关,产生一定时间的滞后。这是产生施加应力和形变之间相位移的原因。将相位移考虑进去,动态测得的模量用实数部分 M' 和虚数部分 M'' 来描述。实数部分(储能模量)描述与周期性应力同相的试样响应,它是试样(可逆的)弹性的量度。虚数部分(损耗模量)描述相位移为 90° 的响应部分,它是转化为热(因而不可逆地损失掉)的机械能量的量度。相位移的正切 $\tan\delta$ 也称作损耗因子,是材料阻尼性能的量度。模量和 $\tan\delta$ 与温度和测试频率有关。

In dynamic mechanical analysis, a mechanical modulus is determined as a function of temperature, frequency and amplitude. A periodically changing force (usually sinusoidal) applied to the sample creates a periodic stress in the sample. The sample reacts to this stress and the instrument measures the corresponding deformation behavior. The mechanical modulus, M, is determined from the stress and deformation. Depending on the type of stress applied, either the shear modulus, G (with shear stress) or the Young's modulus, E (with stretching or bending) is measured. The sample does not always immediately react to the periodically changing stress-a certain time delay occurs that depends on the viscoelastic properties of the sample. This is the cause of the phase shift between the applied stress and the deformation. To take this phase shift into account, the dynamically measured modulus is described by a real part M' and an imaginary part M''. The real part (storage modulus) describes the response of the sample in phase with the periodic stress. It is a measure of the (reversible) elasticity of the sample. The imaginary part (loss modulus) describes the component of the response that is phase-shifted by 90°. This is a measure of mechanical energy converted to heat (and therefore irreversibly lost). The tangent of the phase shift, $\tan\delta$, is also known as the loss factor and is a measure of the damping behavior of the material. The modulus and $\tan\delta$ depend on the temperature and the measuring frequency.

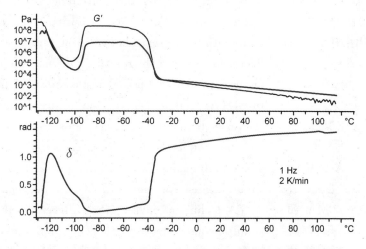

图 1.8　硅油的储能模量 G'、损耗模量 G'' 和损耗角 δ 曲线
1 玻璃化转变(-115℃);2 结晶(-110℃);3 晶体熔融(-40℃);4 液态

Figure 1.8　Curves of the storage modulus, G', and the loss modulus, G'' and the phase shift, δ, as a function of temperature for a silicone oil.
1 glass transition (-115℃); 2 crystallization (-110℃); 3 melting of the crystals (-40℃); 4 liquid state.

图 1.8 所示曲线为硅油在玻璃化转变(1)、结晶(2)和熔融(3)过程和处于液态(4)时储能模量、损耗模量和相角的变化。

The curves in Figure 1.8 show the changes of the storage and loss modulus and the phase shift of a silicone oil during the process of glass transition (1), crystallization (2) and melting (3) and at the liquid state (4).

1.5 与 TGA 的同步测量 Simultaneous measurements with TGA

如 1.2 节所述,可用 TGA 极其灵敏地测量蒸发或反应的质量损失。然而,测量结果的解释常常需要附加的信息。这可以将两个或两个以上技术组合成一个仪器系统得以实现。常用技术为同步 DTA 或同步 DSC 和各种在线气体分析方法(例如质谱和红外光谱)。

As already mentioned in Section 1.2, the loss of mass due to vaporization or in reactions can be detected with great sensitivity using TGA. The interpretation of the measurement results, however, frequently requires additional information. This can be obtained by combining two or more suitable techniques into one instrument system. Techniques often used are simultaneous DTA or simultaneous DSC and various methods for on-line gas analysis (e.g. mass spectrometry and infrared spectroscopy).

1.5.1 同步 DSC 和差热分析(DTA、SDTA)
Simultaneous DSC and Differential thermal analysis

同步 DSC 的测量原理和功能与 1.1.1 节中提到的常规 DSC 是一样的。差热分析(DTA)是将试样和参比物在炉体中升温,试样和参比物之间的温度差用热电偶测量。如果试样中发生热效应(例如相转变或化学反应),额外的能量吸收或释放会改变试样的升温速率。这导致试样和参比两边之间的温差。例如,在放热反应中,试样和参比的温差大于反应前后。热效应用出现的台阶和峰来表示,正如 DSC 曲线。

The measuring principle and functions of simultaneous DSC are same as those of a conventional DSC mentioned in Section 1.1.1. In differential thermal analysis (DTA), a sample and a reference material are heated in a furnace. The temperature difference between the sample and the reference material is measured using thermocouples. If a thermal event occurs in the sample (such as a phase transition or chemical reaction), the additional uptake or release of energy changes the heating rate of the sample. This results in a temperature difference between the sample and reference sides. For example, during an exothermic reaction, the temperature difference between the sample and reference is larger than before or after the reaction. Thermal effects are indicated by the presence of steps and peaks, just as in a DSC measurement curve.

在 SDTA(单式 DTA)中,未用参比样。程序温度相当于参比温度,试样温度是直接测量的。

In SDTA (single DTA), no reference sample is used. The reference temperature corresponds to the program temperature and the sample temperature is measured directly.

同步 DSC 技术可用 TGA 实验同步测量 DSC 曲线,同步 DTA 或 SDTA 技术可在 TGA、TMA 和 DMA 实验中同步测量差热信号。这种信号常常有助于诠释测量结果,因为它可测出并不伴随着质量或尺寸变化的热效应。

The simultaneous DSC technique enables the DSC curve to be simultaneously measured in TGA experiments, and the DTA or SDTA technique enables the differential thermal signals to be simultaneously measured in TGA, TMA and DMA experiments. This often aids interpretation because it detects thermal events that are not accompanied by a change in mass or dimensions.

1.5.2 逸出气体分析（EGA） Evolved gas analysis

为了可靠解释 TGA 曲线,常常要更多地了解 TGA 中试样逸出气体的性质。可将 TGA 仪器用加热输送管连接到气体分析仪来获得这些信息。可几乎同步地分析气体。最常用的两种在线联用是 TGA/MS（一台 TGA 耦联到一台质谱仪 MS）和 TGA/FTIR（一台 TGA 耦联到一台傅立叶变换红外光谱仪 FTIR）。
这些技术的使用细节和应用可参阅《热分析应用手册》的《逸出气体分析》分册。

For a reliable interpretation of TGA curves, one would often like to know more about the nature of the gases evolved from the sample in the TGA. This information can be obtained by connecting the TGA instrument to a gas analyzer by means of a heated transfer line. This allows the gases to be analyzed almost simultaneously. The two most frequently used on-line combinations are TGA-MS (a TGA coupled to a mass spectrometer, MS), or TGA-FTIR (a TGA coupled to a Fourier transform infrared spectrometer, FTIR).
Practical details and applications of these techniques are given in "EGA" of "Application Handbooks Thermal Analysis".

1.5.2.1 TGA/MS

质谱仪（MS）由离子源、分析器和检测器组成。来自 TGA 的气体混合物（挥发性化合物/分解产物和吹扫气体）在离子源中离子化形成分子离子和大量的碎片离子。离子按照质荷比（m/z）在分析器中被分开,然后在检测器系统中被记录。测得的离子谱显示出分子离子和大量从不同分子的碎裂形成的碎片离子。图谱或碎裂方式是所测试的特定化合物的特征。此外,有些元素具有非常特征的同位素波谱,例如氯。
一般来说,可通过测得的质谱的碎裂方式来鉴定和表征逸出气体。测得的图谱还可与图谱数据库中的图谱集对比。
在与 TGA 联用中,并没有必要、通常也不在整个 TGA 测试范围内反复测试（即扫描）全部质量范围。可单独选择特定物质特有的质荷比（m/z）的碎片离子,记录其强度与时间或温度的关系。连续测量有限量特定碎片的这个技术称为多离子检测（MID）或选择离子监测（SIM）,可使灵敏度大大提高。
MS 离子曲线与 TGA（或 DTG）曲线的比较能鉴定或确认热效应中失重的物质的存在。

A mass spectrometer (MS) consists of an ion source, an analyzer and a detector. The gas mixture (volatile compounds/decomposition products and purge gas) arriving from the TGA is ionized in the ion source with the formation of molecular ions and numerous fragment ions. The ions are separated according to their mass-to-charge ratio (m/z) in the analyzer and then recorded by the detector system. The resulting mass spectrum displays the molecular ions and a large number of fragment ions formed from the fragmentation of the different molecules. The spectrum or fragmentation pattern is characteristic of the particular compound measured. Furthermore, some elements have very characteristic isotope patterns, e. g. chlorine.
In general, evolved gases can be identified or characterized by the fragmentation pattern of the mass spectra measured. The spectra obtained can also be compared with collections of spectra in spectral databases.
In combination with a TGA it is not always necessary or usual to measure (i. e. scan) the entire mass range repeatedly at short intervals throughout the TGA measurement. Fragment ions of mass-to-charge (m/z) ratio characteristic for particular substances can be individually selected and their intensities recorded as a function of time or temperature. This technique of continuously measuring a limited number of specific fragments is known as multiple ion detection (MID) or selected ion monitoring (SIM) and results in a large increase in sensitivity.
A comparison of the MS ion curves with the TGA (or DTG) curve allows the presence of a substance responsible for the mass loss in a thermal effect to be identified or confirmed.

1.5.2.2 TGA/FTIR

红外光谱法也能用来鉴定 TGA 分析中形成的气体产物。TGA 仪器通过加热输送管与傅立叶变换红外光谱仪（FTIR）直接耦联，测量自 TGA 仪器到达 FTIR 气体池的气体混合物（挥发性化合物/分解产物和吹扫气体）在 4000 cm^{-1} 至 400 cm^{-1}（波数）范围内的红外图谱。与分子中不同键的特定振动频率对应的波长处的红外能量被吸收，这能鉴定官能团和得到关于有关物质的性质的信息。

现代 FTIR 仪器能在几秒内测定高质量的图谱，在整个 TGA 分析范围内连续记录图谱，这意味着几乎对 TGA 分析的任何温度或时间点都可获得单独谱图，这些谱图可与数据库参比谱图作比较，进行评估。

在实践中，为了简化数据处理，对包含谱图中全部或只是部分信息的与时间（或温度）关系曲线进行记录，这些曲线称为 Gram-Schmidt 曲线或化学谱。

简单说来，Gram-Schmidt 曲线是通过对每个单独图谱的吸收波带强度的积分得到的，该信息表示为与时间关系的吸收强度曲线。曲线是任何瞬间到达 FTIR 光谱仪气体池内的物质量的量度。该曲线一般与 DTG 曲线相似，除了由于气体在连接两个仪器的传输管内输送产生的 DTG 和 Gram-Schmidt 峰之间明显存在的短滞后（通常只有几秒）。因为单独物质的红外吸收强度很不同，所以强度不是直接可比的。

化学谱表示特定波数区域的红外吸收与时间（或温度）的关系，能（通过官能团）观察气体产物的形成与时间（或温度）的关系。随后，检索对应 DTG 峰期间测量的全部范围的 FTIR 图谱，并进行解释。

Infrared spectroscopy can also be used to identify gaseous products formed in a TGA analysis. The TGA instrument is coupled directly to a Fourier transformation infrared spectrometer (FTIR) by a heated transfer line. Infrared spectra of the gas mixture (volatile compounds/decomposition products and purge gas) arriving in the FTIR gas cell from the TGA instrument are measured in the range 4000 cm^{-1} to 400 cm^{-1} (wavenumbers). Infrared energy is absorbed at wavelengths that correspond to the specific vibrational frequencies of the different bonds in the molecules. This allows functional groups to be identified and information on the nature of the substances involved to be gained.

A modern FTIR spectrometer is capable of measuring good quality spectra within a few seconds. The spectra are recorded continuously throughout the TGA analysis. This means that individual spectra are available for practically any temperature or point in time of the TGA analysis. These spectra can be compared with database reference spectra and evaluated.

In practice, to simplify the evaluation, curves are recorded as a function of time (or temperature) that make use of all or just part of the information included in the spectrum. These curves are known as Gram-Schmidt curves and chemigrams.

In simple terms, the Gram-Schmidt curve is obtained by integrating the absorption band intensities of each individual spectrum. The information is presented as a curve of absorption intensity plotted as a function of time. The curve is a measure of the quantity of substance arriving in the gas cell of the FTIR spectrometer at any instant. The curves often resemble DTG curves except that there is obviously a short delay (normally just a few seconds) between the DTG and the Gram-Schmidt peaks due to the gas transport in the transfer line connecting the two instruments. The intensities are not directly comparable because individual substances have very different infrared absorption intensities.

Chemigrams display the infrared absorption of particular wavenumber regions as a function of time (or temperature and allow the formation of gaseous products (via functional groups) to be observed as a function of time. Afterward, the corresponding full range FTIR spectra measured during the DTG peak are recalled and interpreted.

2 热分析在医药工业的应用
Applications of Thermal Analysis in the Pharmaceutical Industry

2.1 热分析药物应用一览表 Application Overview Pharmaceuticals

表格所示为可用热分析测试药物的效应和性能。黑点表示较重要的应用。

The table shows the effects and properties of pharmaceutical products that can be investigated by thermal analysis. The more important applications are marked with a black spot.

	DSC	TGA	TMA	DMA
熔点、熔程 Melting point, melting range	●		○	○
熔融行为、熔融分数 Melting behavior, fraction melted	●		○	○
熔融热 Heat of fusion	●			
纯度 Purity	●	○		
多晶型 Polymorphism	●			
假多晶型 Pseudopolymorphism	●	●		
相图 Phase diagrams	●			
挥发、解吸附、蒸发 Evaporation, desorption, vaporization	○	●		
玻璃化转变 Glass transition	●		●	●
相互作用、相容性 Interaction, compatability	●	○		
热稳定性 Thermal stability	●	●		
氧化稳定性 Oxidation stability	●	○		
分解动力学 Kinetics of decomposition	●	●		
组成分析 Analysis of composition		●		

2.2 制药工业评说 Some Comments on the Pharmaceutical Industry

药剂学是关于药物制剂的性质、作用、制备和配发过程的学科。相比之下,制药工业主要关注药物制剂的开发和生产。

A Pharmacy is the science that concerns itself with the nature, the action, the preparation and delivery processes of pharmaceutical preparations. In contrast, the pharmaceutical industry is concerned mainly with the development and production of pharmaceutical preparations.

药物制剂提供药物活性物质,即药物能够供给身体的途径,从而使与实施途径有关的生理方面(口腔、直肠、皮肤、皮下等)和药的化学物理性能均适合人体。药物制剂由真正的药物即活性成分与辅料(填料、添

Pharmaceutical preparations provide the means by which pharmaceutically active substances or drugs can be supplied to the body, so that both the physiological considerations concerning the means of application (oral, rectal, cutaneous, sub-cutaneous etc.) and the chemical-physical properties of the drug are suitable. A pharmaceutical preparation consists of the actual

加剂等)组成,均须配比正确。

活性组分即药物是供给身体的物质,起治疗作用,或协助治未病(预防),或减轻疾病的后遗症。另一类药是用来诊断疾病的。

药物制剂可依据实施途径(口腔、直肠、皮肤、皮下等)或盖仑配药形式(固态、液态、半液态等)分为不同类型。

因为产品是用来医治人类的,所以应用的质量标准极高。还必须考虑伦理因素。

质量标准一方面包含无菌或高度无菌、应用时的活性证明、无副作用(毒性方面),另一方面要求具有良好的生产规范(GMP)、使用高质量原料、保证在规定使用期限内的产品质量。

为了保证这些质量要求,需要合适的分析方法。必须通过校准调整仪器和确认分析方法来检查和验证分析结果的正确性。

热分析方法可单独使用,也可相互结合使用,此外还可与其他技术(光谱法、色谱法等)联用。

drug(s) or active ingredient(s) together with excipients (fillers, additives etc.), all of which must be present in the right proportions. The active ingredients or drugs are the substances that are supplied to the body and that serve to cure or to assist the preventive treatment (prophylaxe), or to alleviate the results of disease or illness. Another group of drugs is used to diagnose illness.

Pharmaceutical preparations can be divided into different groups depending on their means of application (oral, rectal, dermal, subcutaneous etc.) or their galenic form (solid, liquid, semi-liquid etc.).

Since such products are used to treat human beings, the standards of quality applied are extremely high. Ethical considerations must also be taken into account.

The quality standards include on the one hand sterility or a high degree of sterility, proof of activity at the time of application and absence of undesired effects (toxicological aspects) as well as the demands of good manufacturing practice (GMP), the use of high quality raw materials and the guarantee of the quality of a product during its planned lifetime.

Suitable analytical procedures are required in order to guarantee these quality demands. The correctness of the analytical results must be checked and proven by calibrating and adjusting the instruments and by validating the analytical methods.

Thermoanalytical methods can be used individually, in combination with each other, in simultaneous combination as well as in addition to other techniques (Spectroscopy, Chromatography etc.).

2.3 热分析在药物上的应用
Applications of Thermal Analysis in Pharmaceuticals

多晶型和假多晶型的研究是热分析应用的重要领域。许多药物以几种晶型存在,它们对溶解性和生物利用度有重要影响。其他应用领域包括纯度测定、稳定性测试、物质表征和鉴别、活性成分与非活性成分相互作用的研究、相图测定、无定形物质的玻璃化转变测量等。

热分析在药物行业所研究的药物的最重要效应或性能在下面各小节中叙述。

The investigation of polymorphism and psuedopolymorphism is an important field of application of thermal analysis. Many pharmaceutical substances exist in several crystalline forms, which can have a very important influence on solubility and bio-availability. Other areas of application include purity determination, stability testing, the characterization and identification of substances, the investigation of any possible interaction between the active and inactive ingredients, the determination of phase diagrams and the measurement of the glass transition of amorphous substances.

The most important effects or properties that can be investigated by thermal analysis in the pharmaceutical industry are discussed briefly in the following sections.

2.3.1 多晶型 Polymorphism

多晶型是用来描述物质以不同晶体形式存在能力的术语。即使化学组成相同，但其熔点、熔融热、溶解行为或生物利用度等物理性能会不同。这些不同与物质有关，可以差别很大。例如，一种晶型可以很好吸收，而另一种晶型可能是无效的，甚至是有毒的。因此，了解不同的晶型、能够检测它们、了解它们的转变行为（稳定的/亚稳的）是很重要的，以优化产品及其贮存条件，从而只让所希望的晶型存在。

用 DSC 研究多晶型时，测量参数尤其重要，因为它们会影响不同晶型的转变动力学。对于未知体系，建议先用不同的升温和降温速率进行预测。

Polymorphism is the term used to describe the ability of a substance to exist in different crystalline forms. Even though the chemical composition is the same, different physical properties can result such as melting point, heat of fusion, solubility behavior or bio-availability. These differences are substance dependent and can be quite large. For instance, one polymorphic form can be absorbed much better while the other can be inactive or even toxic. It is therefore important to be aware of the different modifications, to be able to detect them and to have information on their transition behavior (stable/metastable) in order to be able to optimize the production and storage conditions so that only the desired form is present.

In the investigation of polymorphism with DSC, the measurement parameters are especially important because these can influence the transition kinetics of the different modifications. With unknown systems it is recommended that the investigation be first carried out at different heating and cooling rates.

2.3.2 假多晶型 Pseudopolymorphism

假多晶型常用来描述药物或添加剂的水合物或溶剂化物。在制备过程中，当物质结晶析出时有水或溶剂结合在其晶格中，便产生了这类化合物。以这种方式结合进的水或溶剂分子，与只是吸附在晶体表面的分子（例如水分）是不同的，需要在较高的温度去溶剂化。理想的研究方法是联合使用 DSC 和 TGA。

This expression is often used to describe hydrates or solvates of drugs or additives. These compounds are produced when, during its preparation, a substance crystallizes out and incorporates molecules of water or solvent in its crystal lattice. These molecules of water or solvent bound in this way are different to molecules that are merely absorbed on the surface of the crystal (e.g. moisture) in as much as that a higher temperature is required for their desolvation. The ideal method of investigation is a combination of DSC and TGA.

2.3.3 相图 Phase diagrams

在药物制剂开发中的重要问题是组分是否相溶，是否存在相溶间歇，是否形成共熔体。用相图可解答这些问题，相图描述多组分体系熔融温度与组成间的关系。要绘制相图，可用 DSC 测量不同组成的混合物，计算其熔点、熔程等数据。

Important questions in the development of pharmaceutical preparations are whether the components used are miscible, whether there are miscibility gaps or whether a eutectic is formed. These questions can be answered with a phase diagram, which describes the relationship between the melting temperature and composition of a multi-component system. In order to construct such a phase diagram, mixtures of the components with different compositions are measured with DSC and the data evaluated (e.g. melting point, melting range).

2.3.4 稳定性 Stability

稳定性和使药物制剂稳定化的课题越来越重要。毕竟,考虑到贮存和运输的要求,了解产品可保持多久是非常重要的。用长期测试或者借助于动力学测量,可测定稳定性。

长期测试是对不同条件下不同时间的贮存样品按一定间隔进行分析。由测量结果的变化可很容易确认产品状态的变化。

此外,有一个快速方法,通过在不同动态条件(升温速率)下测量试样,动态研究分解反应。可应用不同的计算方法,来确定动力学参数。得到的结果能对分解行为作出某些预测。不过,仅能用于解释趋势分析。

The subject of stability and the means of stabilizing pharmaceutical preparations is a topic of ever increasing significance. It is, after all, very important to know how long a product can be kept, taking into account the requirements of storage and distribution. The stability can be determined with long-term tests or with the help of kinetic measurements.

In the case of long-term tests, samples that have been stored for different periods of time under different conditions are analyzed at regular intervals. Changes in the state of the product can be easily recognized as shift or change of measurement results.

In addition, there is a rapid procedure in which the decomposition reactions are investigated kinetically by measuring the samples under different dynamic conditions (heating rates). Different evaluation methods can be applied in order to determine the kinetic parameters. These results allow certain predictions about the decomposition behavior to be made. They should however only be interpreted in the form of a trend analysis.

2.3.5 相互作用 Interactions

本节中的相互作用一词,意指药物制剂的两个或多个组分间的作用。这种相互作用可以是符合需要的或者不合需要的。符合需要的相互作用是有意使之发生的,为的是改善活性组分的溶解性。不合需要的相互作用,又称作不相容性,会造成一个或多个组分的改变,从而使得制剂失去活性甚至完全失效。在最差情况下,甚至会产生有毒分解产物。用 DSC 或 TGA,通过对各个组分和混合物分析结果的相互比较,可容易地研究药物制剂。如果混合物呈现各个组分不明显存在的热效应,则表示有相互作用。不过,对结果的解释必须非常谨慎:例如,共熔体系呈现并不是由不相容性产生的热变化。此外,在加热的开始阶段,存在物质变化的危险(例如水的蒸发)。因此,不能用数据外推来预测

The term interaction, in this context, means interaction between two or more components of a pharmaceutical preparation. Such interactions can be desirable or undesirable. Desirable interactions are purposely brought about in order to improve the solubility of an active ingredient. Undesirable interactions, also known as incompatibility, are interactions that lead to changes in one or more of the components which then bring about a loss of activity or even a complete deactivation of a preparation. Even toxic decomposition products can be produced in the most unfavorable cases.

Pharmaceutical preparations can easily be investigated by DSC or TGA by comparing the results for the individual components and the mixture with each other. If the mixture exhibits thermal effects that are not apparent in the individual components then this is an indication of an interaction. However the results must be interpreted very carefully: a eutectic system, for example, shows thermal changes which are not caused by incompatability. In addition, there is the danger that the substance changes during the initial heating stage (e.g. evaporation of water). An extrapolation of data to predict behavior at lower temperatures is

低温下的行为。这时,较好的做法是,把试样在恒温和确定的湿度下储存,每隔一定时间用DSC进行分析。

therefore not possible. It is better, in such cases, to store the samples at constant temperature and with defined humidity, and to analyze them at regular intervals with DSC.

2.3.6 纯度测定 Purity determination

制备药品时,使用纯的活性和非活性组分是绝对必要的,因为不合需要的杂质可能会造成十分严重的后果。因此,日常须进行纯度测定,测定方法是依据观察有机化合物存在杂质会使熔点降低,即熔点随着杂质量的增加而降低。对于稀释体系,熔点降低与杂质含量的关系由van't Hoff方程表示。严格来说,当满足下列条件时此式才成立:存在共熔体系,组分在液态是相溶的,存在热力学平衡,熔融时不发生分解。

考虑到这些限制,尤其重要的是,该方法对于经过充分研究的体系才适合进行常规样品批量分析。对单次DSC测试的熔融曲线的分析,就可得到测定物质纯度、熔点和熔融热所要求的所有信息。

The use of pure active and inactive ingredients for the preparation of pharmaceutical products is absolutely essential, since undesirable impurities could have very serious consequences. For this reason the determination of purity is performed routinely. The method is based on the observation that the presence of impurities in an organic compound depresses the melting point i. e. the melting point is lowered with increasing amounts of impurity. The correlation between the melting point depression and the degree of impurity for dilute systems is described by the van't Hoff equation. Strictly speaking, this is valid only when the following conditions are fulfilled: a eutectic system is present, the components are miscible in the liquid state, a thermodynamic equilibrium is present and no decomposition occurs on melting.

Taking these limitations into account, the method is above all suitable for the routine analysis of sample lots with systems that have been well investigated. The analysis of the melting curve of a single DSC measurement yields all the information required to determine the purity, the melting point and the heat of fusion of a substance.

2.3.7 包装材料 Packaging materials

各种包装材料在制药工业中也是很重要的。必须区分与药物制剂直接接触的包装材料和仅用作外包装的包装材料。对前类包装材料有十分严格的要求,因为有时可能与药物制剂发生反应。因此,合适材料的选择和鉴定十分重要。由于合成聚合物材料越来越多地用作包装材料,所以也将热分析用于质量控制和鉴定。

Packaging materials of all types are also important in the pharmaceutical industry. One must distinguish between packaging that is in direct contact with the pharmaceutical preparation and packaging that serves only as external packaging. Very stringent requirements apply for the first type of packaging because it could in some instances react with the pharmaceutical preparation. The choice of a suitable material and its identification is therefore of great importance. Because synthetic polymers are increasingly being used as packaging material, thermal analysis is also being employed for the quality control and identification purposes.

2.3.8 工艺优化 Process optimization

用热分析测量的效应(例如熔融)也

Effects that can be measured with thermoanalysis (e. g. melting)

用来直接表示物质是否在过程中经历变化,因而,可能的话,必须选择工艺参数使之不发生变化。

are also used to show indirectly whether a substance suffers a change during processing. The process parameters must then be chosen so that, if possible, this does not happen.

2.3.9　校准和系统效应　Calibration, systematic effects

由于对分析方法的质量所要求的标准高,所以方法参数(测试条件、样品制备)的准确性(校准)及其影响特别重要。因此,须注意方法的影响和实验参数的选择。

Because of the high standards required for the quality of analytical procedures, special importance is attached to the accuracy (calibration) and influence of method parameters (measurement conditions, sample preparation). Attention is therefore paid to the effects of methodology and the choice of experimental parameters.

2.3.10　一些重要概念和缩写　Some important Concepts, Abbreviations and Acronyms

调整:通过改变仪器参数使测量指示值落在规定误差限之内来使仪器性能最优化。
ASA:乙酰水杨酸。
BHA:丁基羟基茴香醚。
校准:由法定的或法律认可的个人或职权者对(仪器)设定的检查;或者是指示值与用合格的参比物质测得的真实值之间关系的测定。

Conc:浓度。
c_p:比热容。
交叉污染:药物被少量杂质污染,特别是在生产过程中,例如使用未被合格清洗的生产机器。交叉污染可能是均相分布的,也可能是非均相分布的。取决于污染物质的作用,可能会对病人产生灾难性后果。

D:不对称 C-原子构象的符号。
DSC:差示扫描量热法。
DTG:微商热重分析法(TGA 曲线的一阶微商)。
EGA:逸出气体分析(联用技术举例:TGA 与 MS 或 FTIR)。
FTIR:傅里叶变换红外光谱。
GLP:关于组织方法和条件的良好实验室规范,据此进行实验室测试的计划、进行和监测以及测试结果

Adjustment: Optimizing the performance of an instrument by making changes to the instrument so that the indicated measurement value lies within the specified limits of error.
ASA: Acetyl salicylic acid.
BHA: Butylated hydroxyanisole.
Calibration: Calibration or checking of the (instrument) settings by an official or an officially recognized person or authority, or: the determination of the relationship between the indicated value and the true value of the measured quantity using certified reference materials.
Conc: Concentration.
c_p: Specific heat capacity.
Cross-Contamination: Contamination of a pharmaceutical substance with a small amount of a foreign substance especially during production for instance by using a production machine that had not been adequately cleaned. Such cross-contamination can be homogeniously or inhomogeniously distributed. Depending on the action of the contaminant substance, it can have disastrous consequences for the patient.

D: Symbol for the configuration of an assymetric C-atom.
DSC: Differential Scanning Calorimetry.
DTG: Derivative Thermogravimetry (first derivative of the TGA curve).
EGA: Evolved Gas Analysis (combination technique e.g. TGA with MS or FTIR).
FTIR: Fourier Transform Infrarot Spectroscopy.
GLP: Good Laboratory Practice concerns itself with organizational procedures and conditions, under which laboratory tests are planned, performed and monitored as well as the recording and

的记录和归档。

GMP：良好生产规范，涵盖药物一致配料生产的要求和建议，以使符合关于活性成分特性、纯度、含量和释放的指定指标。各国的 GMP 要求可以不相同。

ΔH：熔融、反应等的比焓或单位质量的热量变化(J/g)。

HPLC：高压（即高效）液相色谱。

HPMC：羟基丙氧基甲基纤维素。

HSM：热台显微镜法。

ICTAC：国际热分析和量热学协会。

IPC：在线过程控制：在可能影响产品质量的所有关键步骤的生产中进行测量和系统控制。

IUPAC：国际理论与应用化学联合会。

L：不对称 C-原子构象的符号。

m/e：（质谱中）离子的质/荷比。

MpHB：对羟基苯甲酸甲酯，4-羟基苯甲酸甲酯的旧名。

pHB：对羟基苯甲酸，4-羟基苯甲酸的旧名。

MID：多离子检测（质谱分析法中的检测技术）。

MS：质谱分析法。

Na-CMC：羧甲基纤维素钠。

NBS：国家标准局（美国）。

O/W：水中油。

PABS：对氨基苯甲酸，4-氨基苯甲酸的旧名。

PE-HD：高密度聚乙烯（旧名 HDPE）。

PEG：聚乙二醇。

PE-LD：低密度聚乙烯（旧名 LDPE）。

PE-LLD：线性低密度聚乙烯（旧名 LLDPE）。

PS：聚苯乙烯。

P_2O_5：五氧化二磷（干燥剂）。

QC/QA：质量控制（QC）以及质量保证（QA）包含与药物及其产品的特性、纯度、组成和其他性能有关的

documentation of the results of tests.

GMP：Good Manufacturing Practice, covers the requirements and recommendations for the production of uniform charges of pharmaceutical products so that they conform to the prescribed specifications regarding identity, purity, content and release of the active ingredient. GMP requirements can vary from country to country.

ΔH：Specific change of enthalpy or heat of fusion, reaction etc J/g.

HPLC：High Presure (or High Performance) Liquid Chromatography.

HPMC：Hydroxypropoxymethylcellulose.

HSM：Hot Stage Microscopy = Thermomicroscopy.

ICTAC：International Confederation for Thermal Analysis and Calorimetry.

IPC：In-Process-Control：Measurement and systematic control during production of all the critical steps that can influence the quality of a product.

IUPAC：International Union of Pure and Applied Chemistry.

L：Symbol for the configuration of an assymetric C-atom.

m/e：Mass to charge ratio of an ion (in mass spectrometry).

MpHB：Methyl p-hydroxybenzoate, old name for 4-Hydroxybenzoic acid methyl ester.

pHB：p-Hydroxybenzoic acid, old name for 4-Hydroxybenzoic acid.

MID：Multiple Ion Detection (detection technique in mass spectrometry).

MS：Mass spectrometry.

Na-CMC：Sodium carboxymethylcellulose.

NBS：National Bureau of Standards (USA).

O/W：Oil in water.

PABS：P-Aminobenzoic acid, old name for 4-Aminobenzoic acid.

PE-HD：High density polyethylene (formerly named HDPE).

PEG：Polyethyleneglycol.

PE-LD：Low density polyethylene (formerly named LDPE).

PE-LLD：Linear low density polyethylene (formerly named LLDPE).

PS：Polystyrene.

P_2O_5：Phosphorus pentoxide (drying agent).

QC/QA：The quality control (QC) resp. quality assurance (QA) includes all activities that are concerned with the control of identity, of purity, of composition and of other properties of a

一切活动。

参比物：经常称为标准物，呈现一个或若干个确定的化学或物理效应的均相、稳定的物质。

r. h.：相对湿度。

SDTA：单盘差热分析。

DTA：差热分析。

TBA：甲苯磺丁脲。

T_f：熔融温度。

T_g：玻璃化转变温度。

TGA：热重分析。

USP：美国药典。

确认：所有必要设备和生产与分析步骤的系统测试和文件归档，以保证特定产品符合需要的质量。

（＋）、（－）：光学活性物质光旋转的方向。

pharmaceutical substance or product.

Reference: Is often referred to as a Standard. Is a homogenous, stable material that exhibits one or more defined chemical or physical effects.

r. h.: relative humidity.

SDTA: Single Differential Thermal Analysis.

DTA: Differential Thermal Analysis.

TBA: Tolbutamide.

T_f: Temperature of fusion.

T_g: Glass transition temperature.

TGA: Thermogravimetric analysis, often referred to as TG.

USP: United States Pharmacopeia.

Validation: The systematic testing and documentation of all the essential equipment and steps in production and analysis in order to guaranty the desired quality of a particular product.

（＋），（－）: Direction of rotation of light for optically active substances.

3 热分析的药物典型应用
Typical TA Applications of Pharmaceuticals

3.1 DSC 温度和热流量的校准
DSC Calibration, Temperature and Heat flow

样品	铟(校准标准物,纯度>99.999%)		
Sample	Indium (calibration standard, purity > 99.999%)		
应用	校准标准物		
Application	Standard for calibration		
条件	测试仪器:DSC	**Measuring cell**: DSC
Conditions	坩埚:40 μl 铝坩埚,盖钻孔	**Pan**: Al 40 μl, with pierced lid
	样品制备:将铟粒压平,预熔	**Sample preparation**: Indium pellet, pressed flat, premelted
	测试:	**Measurement**:
	以 10 K/min 由 120 ℃升温至 180 ℃	Heating from 120 ℃ to 180 ℃ at 10 K/min
	气氛:氮气,50 ml/min	**Atmosphere**: Nitrogen, 50 ml/min

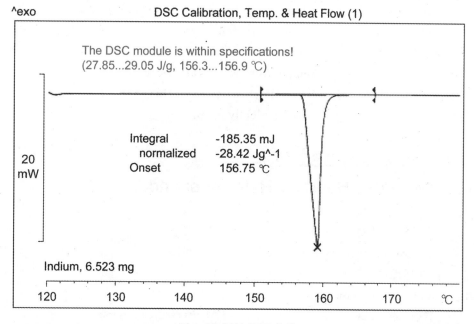

图 3.1 铟的 DSC 曲线
Figure 3.1 DSC curve of indium

解释 图 3.1 为铟熔融的 DSC 曲线。纯物质在完全确定的温度即熔点熔融。取熔融过程的始点(或称外推始点)为熔点,定义为由熔融曲线外推斜率与基线延长线交点的温度。

Interpretation The DSC curve in Figure 3.1 shows the melting of indium. A pure substance melts at an exactly defined temperature, its melting point. The melting point is taken to be the start or onset of the melting process which is defined as the temperature given by the intercept of the extrapolated slope of the melting curve and the continuation of the base line.

计算 测定了铟的熔融起始温度和熔融热,全自动的评估会进行确认,即将测量值与文献值做比较。如果比较值落在允许误差范围内(如本例所示),则在图上显示"DSC 仪器符合指标"的信息。

Evaluation The onset temperature and the heat of fusion of indium are determined. The fully automated evaluation performs a validation which compares the measured values with literature values. If, as in this case, the values lie within the allowed limits then the message 'The DSC module is within specifications' is displayed as in the figure.

	测量值 Measured	标准值 Ref. value	允许误差 Tolerance
熔点(外推始点) Melting point(onset) ℃	156.75	156.60	±0.3
熔融热 Heat of fusion J/g	28.42	28.45	±0.6J/g

结论 所谓"铟检查"是检查仪器温度和热流量校准的快捷方法。测量结果与标准值自动比较,如果仪器需要调整,则仪器会显示相应的信息。

如果仪器经常用于其他温度范围,则建议用适合这些温度范围的另外的标准物质做进一步检查。

本例中给出的允许误差为标准值,每个人可自行修改。

Conclusion The so-called "indium-check" is a quick and easy method to check the temperature and heat flow calibration of an instrument. The results are automatically compared with reference values. The instrument displays the appropriate message if an adjustment of the instrument is required.

If the instrument is frequently used in other temperature ranges, then further checks with additional standards suitable for those temperature ranges are recommended.

The tolerances given in this example are standard values and can be individually adapted.

3.2 与升温速率无关的 DSC 校准
DSC Calibration, Hearting Rate Independence

样品 Sample	锌(校准标准物,纯度>99.999%) Zinc (calibration standard, purity > 99.999%)
应用 Application	校准标准物 Standard for calibration
条件 Conditions	测试仪器:DSC 坩埚:40 μl 铝坩埚,盖钻孔 样品制备:将锌粒预熔 测试: 以 5 K/min、10 K/min 和 20 K/min 由 350 ℃升温至 475 ℃。所有测试用同一样品。降温速率为 5 K/min。 气氛:氮气,50 ml/min
	Measuring cell:DSC **Pan**:Al 40 μl, with pierced lid **Sample preparation**:Zinc pellet, premelted **Measurement**: Heating from 350 ℃ to 475 ℃ at 5, 10 K/min and 20 K/min. All measurements are performed with the same sample. Cooling rate is 5 K/min. **Atmosphere**:Nitrogen, 50 ml/min

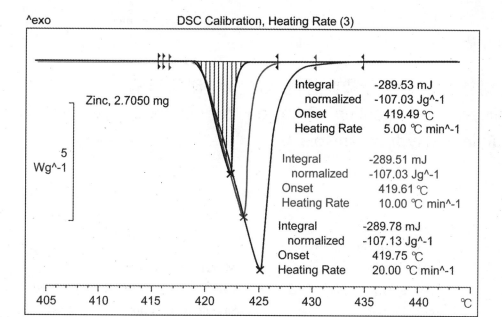

图 3.2 以不同升温速率测量的锌的 DSC 曲线
Figure 3.2 DSC curves of zinc measured at different heating rates

解释 图 3.2 中所示为不同升温速率的 DSC 熔融曲线。如果表示为与温度的关系,则峰面积随着升温速率加快而增大。但熔融热为热流量对时间的积分,如同熔点一样是与升温速率无关的。

Interpretation The DSC curves in Figure 3.2 show the melting of zinc at different heating rates. If displayed with respect to temperature, the peak area increases with increasing heating rates. The heat of fusion is however the integral of the heat flow with respect to time. This is just as independent of the heating rate as the melting point.

计算 测定了锌的熔融起始温度和熔融热。

Evaluation The onset temperature and the heat of fusion of zinc are determined.

	测量值 Measured	测量值 Measured	测量值 Measured	标准值 Ref. value
升温速率 Heating rate K/min	5	10	20	
熔点(起始点) Melting point(onset) ℃	419.5	419.6	419.8	419.7
熔融热 Heat of fusion J/g	107.0	107.0	107.1	107.0

结论 仪器校准受许多因素的影响,例如升温速率、所用的吹扫气体、试样坩埚材料和测试采用的温度范围。只有当将校准数据考虑这些影响因素时,才能获得与测试参

Conclusion The calibration data of an instrument are affected by many factors such as the heating rate, the purge gas used, the sample pan material and the temperature range used. Only when these effects can be taken into account in the calibration data, are results obtained that are independent of measurement parameters, as

数无关的结果,如同本例所示熔点和熔融热的测量结果。

is shown in this example of the melting point and the heat of fusion.

3.3 升温速率对丁基羟基茴香醚多晶型检测的影响
Influence of Heating Rate on the Detection of Polymorphism, Butylated Hydroxyanisole

样品 **Sample**	丁基羟基茴香醚 Butylated hydroxyanisole		
应用 **Application**	非活性成分(抗氧化剂) Inactive ingredient (antioxidant)		
条件 **Conditions**	测试仪器:DSC	**Measuring cell**:DSC	
	坩埚:40 μl 铝坩埚,密封	**Pan**:Al 40 μl, hermetically sealed	
	样品制备:原样品	**Sample preparation**:As received	
	测试:	**Measurement**:	
	以 1 K/min、2.5 K/min、5 K/min、10 K/min 和 20 K/min 由 30 ℃升温至 70 ℃	Heating from 30 ℃ to 70 ℃ at 1, 2.5, 5, 10 and 20 K/min	
	气氛:氮气,50 ml/min	**Atmosphere**:Nitrogen, 50 ml/min	

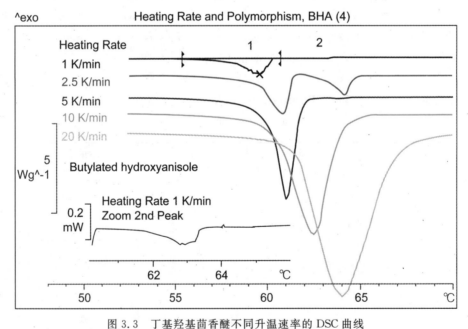

图 3.3 丁基羟基茴香醚不同升温速率的 DSC 曲线
Figure 3.3 DSC curves of butylated hydroxyanisole at different heating rates

解释 以不同升温速率测试的丁基羟基茴香醚熔融的 DSC 曲线如图 3.3 所示。如同预料的,峰的大小随升温速率的加快而增大,结果造成分辨率下降。仅在较低的升温速率观察到第 2 熔融峰,该峰对应于

Interpretation The DSC curves shown in Fig. 3.3 display the melting of butylated hydroxyanisole recorded at different heating rates. As expected the peak increases in size with increasing heating rates, which results in decreasing resolution. A second melting peak, which corresponds to the other modification of butylated hydroxyanisole, is observed only at lower heating

丁基羟基茴香醚另一晶型。从一种晶型转变到另一种晶型的动力学，是多晶型转变的一个重要因素。在本例中,检测第2个转变的最佳升温速率为2.5 K/min。而测量第1种晶体熔融热的最佳速率是10 K/min。需要另外的测试来测定第2种晶型的熔融热。

rates. An additional factor of importance in the case of polymorphic transitions is the kinetics for transformation from one modification to the other. In the case described, the most favorable heating rate for the detection of the second transition is 2.5 K/min. The heat of fusion of the first crystal modification is best measured at 10 K/min. Further experiments are required to determine the heat of fusion of the second crystal modification.

计算
Evaluation

升温速率 Heating rate K/min	起始点1 Onset 1 ℃	ΔH J/g	起始点2 Onset 2 ℃	ΔH J/g	重量 Weight mg
1	59.2	95.0	62.6	2.7	1.360
2.5	59.9	82.1	63.2	17.6	4.929
5	59.4	85.5	62.9	2.9	1.607
10	60.2	100.2	—		4.516
20	61.1	99.3	—		4.398

结论 用合适的升温速率,可测定多晶型物质各个晶型的熔点和熔融热,即使各个晶型的熔点靠的很近。

Conclusion With a suitable choice of the heating rates, it is possible to determine the melting point and the heat of fusion of the individual modifications of polymorphic substances, even when the melting points of the modifications lie close together.

3.4 降温速率对蔗糖溶液结晶行为的影响
Influences of Cooling rate on Crystallization Behavior, Saccharose Solutions

样品 **Sample**	D(+)蔗糖,20％重量水溶液(＝1.05％摩尔) D(+) Saccharose solution, 20 weight％ in water (＝1.05 mole％)
应用 **Application**	非活性成分(溶液稳定剂) Inactive ingredient (solution stabilizer)
条件 **Conditions**	测试仪器：DSC　　**Measuring cell**：DSC 坩埚：40 μl 铝坩埚,密封　　**Pan**：Al 40 μl, hermetically sealed 样品制备：　　**Sample preparation**： 用移液管将2.260 mg重的液滴加入坩埚。　　One drop of solution is weighed into the pan using a pipette, sample weight 2.260 mg. 测试：　　**Measurement**： 以－1 K/min、－2 K/min、－5 K/min、－10 K/min、－20 K/min 由25 ℃降温至－50 ℃。同一试样用于所有测试。以5 K/min 由－50 ℃升温至25 ℃。　　Cooling from 25 ℃ to －50 ℃ at －1, －2, －5, －10, －20 K/min. The same sample is used for all the measurements. Heating from －50 ℃ to 25 ℃ at 5 K/min. 气氛：氮气,80 ml/min　　**Atmosphere**：Nitrogen, 80 ml/min

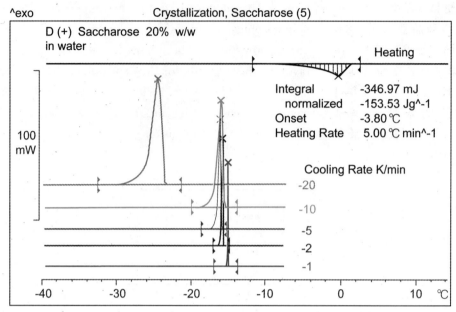

图 3.4　D(＋)蔗糖溶液结晶和熔融过程的 DSC 曲线

Figure 3.4　DSC curves of crystallization and melting processes of the D(＋) Saccharose solution

解释　图 3.4 所示曲线为用 DSC 测试的 D(＋)蔗糖溶液的结晶和熔融过程。降温速率慢时,起始温度几乎不变,但降温速率较快时,起始点降低(过冷)。降温速率很快时,溶液甚至不结晶,而是发生玻璃化,即转变为玻璃态。

由熔融峰可计算得到熔点下降和水的含量。

Interpretation　The curves in Fig. 3.4 show that crystallization and melting processes of the D(＋) Saccharose solution can be measured with the DSC. At low cooling rates, the onset temperatures are almost constant, but are displaced to lower values (supercooling) at higher cooling rates. At very high cooling rates it is even possible that the solution does not crystallize but vitrifies i.e. is transformed to glassy state.

The melting point depression and the 'purity' of the water can be calculated from the melting peak.

计算　用熔融峰算得的纯度:99.02%(理论值:98.95%)

熔点下降:－1.76 ℃

Evaluation　Purity calculated using the Melting peak:99.02 mole% (theoretical value:98.95)

Melting point depression:－1.76 ℃

降温速率 Cooling rate K/min	起始点 Onset ℃	ΔH J/g	效应 Effect
－1	－15.0	173.2	结晶 crystallization
－1	－15.3	168.3	结晶 crystallization
－5	－15.8	162.5	结晶 crystallization
－10	－15.3	170.0	结晶 crystallization
－20	－23.0	157.3	结晶 cristallization
＋5	－3	153.3	熔融 melting

结论　熔融与结晶过程的起始温度是不同的。结晶过程由成核动力学控制,与降温速率和试样量(存在的

Conclusion　The onset temperatures of the melting and crystallization processes are different. Crystallization processes are controlled kinetically by nucleation and are dependent on the cooling rate

晶核数量)有关。熔融峰的起始温度通常不受干扰影响。

and the amount of sample (number of nuclei present). The onset temperatures of melting peaks are not normally subject to disturbing influences.

3.5 升温速率对水包油乳膏水分含量测定的影响
Influences of Heating Rate on Moisture Content Determination, an O/W Cream

样品	647-A 水包油乳膏品		
Sample	O/W Cream sample 647-A		
应用	制造奶油的原料		
Application	Basic material for the manufacture of creams		
条件	测试仪器：TGA	**Measuring cell**：TGA	
Conditions	坩埚：100 μl 铝坩埚，盖钻孔	**Pan**：Al 100 μl, with a pierced lid	
	样品制备：原样品	**Sample preparation**：As received	
	测试：	**Measurement**：	
	1. 以 20 K/min 由 30 ℃升温至 300 ℃。	1. Heating from 30 ℃ to 300 ℃ at 20 K/min.	
	2. 以 5 K/min 由 30 ℃升温至 300 ℃。	2. Heating from 30 ℃ to 300 ℃ at 5 K/min.	
	气氛：氮气,50 ml/min	**Atmosphere**：Nitrogen, 50 ml/min	

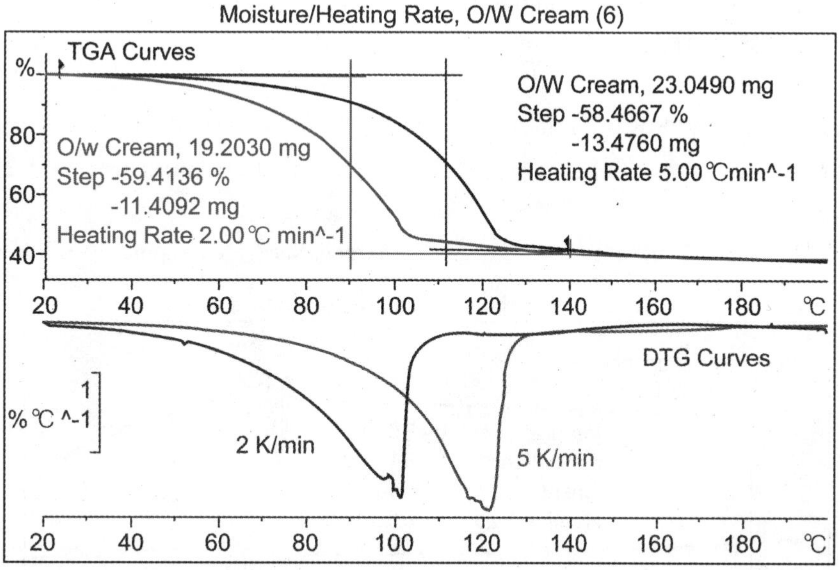

图 3.5　水包油乳膏试样的 TGA 和 DTG 曲线
Figure 3.5　TGA and DTG curves of the O/W Cream sample

解释　图 3.5 中的 TGA 曲线呈现水包油乳膏可挥发成分(主要为水)在 40 ℃至 140 ℃范围内的蒸发。升温速率较高时,蒸发移至较高温度。TGA 曲线的一阶微商 DTG 有助于确定 TGA 信号的最终台阶。

Interpretation　The TGA curves in Fig. 3.5 show the evaporation of the volatile components (mainly water) of O/W Cream in the region between 40 ℃ and 140 ℃. At higher heating rates the evaporation is diplaced to higher temperatures. The first derivative of the TGA curve, DTG, is helpful for the determination of the final step of the TGA signal.

计算 Evaluation	升温速率 Heating rate K/min	台阶 Step %	DTG 峰温 Peak Temperature ℃
	2	59.4	102.1
	5	58.5	122.4

结论 开发方法时已经考虑了试样质量和形状、升温速率以及坩埚类型的影响。当研究与时间和温度有关的效应（例如蒸发）时，升温速率特别重要。

Conclusion The influence of the mass and form of the sample, the heating rate and the type of pan have to be considered when developing methods. The heating rate is of special importance when investigating time and temperature dependent effects such as evaporation.

3.6 升温速率对美托拉腙分解的影响
Influences of Heating Rate on Decomposition, Metolazone

样品 Sample	美托拉腙 Metolazone
应用 Application	活性成分（利尿剂） Active ingredient (diuretic)
条件 Conditions	测试仪器：DSC 坩埚： 40 μl 铝坩埚，盖钻直径 1 mm 的孔。 样品制备：原样品 测试： 以 2 K/min 和 5 K/min 由 20 ℃升温至 200 ℃。两次测试均进行空白修正。 气氛：氮气，50 ml/min

Measuring cell: DSC
Pan: Al 40 μl, with pierced lid with a hole of 1 mm diameter.
Sample preparation: As received
Measurement:
Heating from 20 ℃ to 200 ℃ at 2 K/min and 5 K/min. Both measurements are blank curve corrected.
Atmosphere: Nitrogen, 50 ml/min

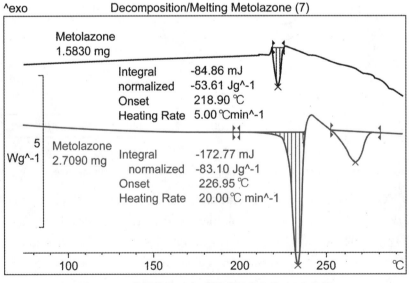

图 3.6　α-美托拉腙两个不同升温速率的 DSC 曲线

Figure 3.6　DSC curves of α-Metolazone measured at two different heating rates

解释 图 3.6 为以两个不同升温速率测量的 α-美托拉腙的 DSC 曲线的比较。以 20 K/min，可观察到 α 晶型在约 227 ℃熔融，随后是 γ 晶型的重结晶，在约 255 ℃该晶型熔化，且伴有分解。

对于升温速率 5 K/min，观察到一个小得多的吸热效应，熔点降低了大约 7 ℃。在 220 ℃，密封坩埚开裂（表现为基线偏移）。

计算 仅计算吸热峰。选择"左水平"或"右水平"基线进行峰积分。

Interpretation The DSC curves of α-Metolazone measured at two different heating rates are compared in Fig. 3.6. At 20 K/min the melting of the α modification at approx. 227 ℃ followed by recrystallization in the γ modification can be observed. This melts with accompanying decomposition at approx. 255 ℃.

With a heating rate 5 K/min, a much smaller endothermic effect and a depression of the melting point by approx. 7 ℃ is observed. At 220 ℃ the hermetically sealed pan bursts (shown as a displacement of the base line).

Evaluation Only the endothermic peaks are evaluated. To integrate the peaks the baseline types 'horizontal left' or 'horizontal right' are selected.

升温速率 Heating rate K/min	起始点 1 Onset 1 ℃	ΔH J/g	起始点 2 Onset 2 ℃	ΔH J/g
5	218.9	53.2	—	—
20	227.0	83.1	255.1	31.0

结论 DSC 对在熔点处分解的物质只能有限表征。这时，用较快的升温速率常常可获得再现性较好的表征，因为分解反应移至较高温度。相同的方法也可用于 TGA，以得到更多的信息，尤其是当生成挥发性分解产物时。

Conclusion Substances which decompose at the melting point can only characterized to a limited extent by DSC. In such cases, a more reproducible characterization is often possible by using faster heating rates, since the decomposition reaction is moved to higher temperatures. TGA can also be used in the same way to gain additional information, especially when volatile decomposition products are formed.

3.7 坩埚对一水葡萄糖失水的影响
Influence of the Pan on Dehydration, Glucose Monohydrate

样品	α-D-一水葡萄糖	
Sample	α-D-Glucose monohydrate	
应用	非活性成分，药片和胶囊的填料	
Application	Inactive ingredient, filler for tablets and capsules	
条件 Conditions	测试仪器：DSC 和 TGA DSC 坩埚：	**Measuring cell**：DSC and TGA **Pan DSC**：
	40 μl 铝坩埚，密封或盖钻孔。	Al 40 μl, hermetically sealed or with pierced lid.
	TGA 坩埚：100 μl 铝坩埚，盖钻孔。	**Pan TGA**：Al 100 μl, with pierced lid.
	样品制备：原样品	**Sample preparation**：As received
	DSC 测试：	**DSC Measurement**：
	以 20 K/min 由 30 ℃升温至 250 ℃。	Heating from 30 ℃ to 250 ℃ at 20 K/min.

TGA 测试：
以 20 K/min 由 30 ℃升温至 300 ℃。
气氛：氮气，DSC：50 ml/min，TGA：80 ml/min

TGA Measurement：
Heating from 30 ℃ to 300 ℃ at 20 K/min.
Atmosphere：
Nitrogen，DSC：50 ml/min，TGA：80 ml/min

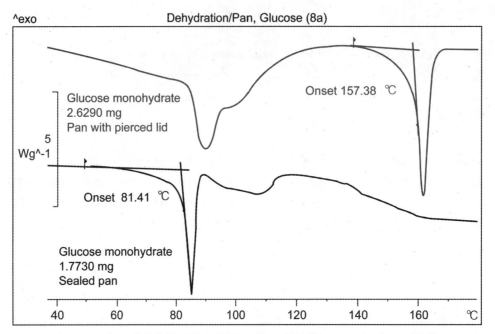

图 3.7　α-D-一水葡萄糖用钻孔盖坩埚（上）和密封坩埚（下）测试的 DSC 曲线
Figure 3.7　DSC curves of α-D-Glucose monohydrate measured at pan with pierced lid (above) and sealed pan (below)

解释　图 3.7 所示 α-D-一水葡萄糖两条 DSC 曲线的比较，表明了试样在密封坩埚或在盖钻孔的坩埚中测试时产生的变化。如果用盖钻孔的坩埚，结晶水可以逃逸，这从测试开始时 DSC 曲线的移动和宽泛的蒸发峰可明显观察到。同时，出现向 α-D-无水葡萄糖的转变，它的熔点约为 158 ℃。200 ℃以上，葡萄糖开始熔化变焦。

Interpretation　A comparison of the two DSC curves of α-D-Glucose monohydrate in Fig. 3.7 shows the changes that arise when the sample is measured in a sealed pan or in a pan with a pierced lid. In a hermetically sealed pan the sharp melting peak of the monohydrate can be observed. If a pan with a pierced lid is used, the water of crystallization can escape. This is noticeable as a shift of the DSC curve at the beginning of the measurement and as a broad evaporation peak. At the same time, a transition to α-D-Glucose anhydrate occurs, the melting point of which is at about 158 ℃. Above 200 ℃ the glucose starts to caramelize.

DSC 计算
Evaluation DSC

测试条件 Measuring conditions	起始点 Onset ℃	效应 Effect
密封坩埚 Sealed pan	81.4	熔融
盖钻孔的坩埚 Pan with pierced lid	157.4	熔融

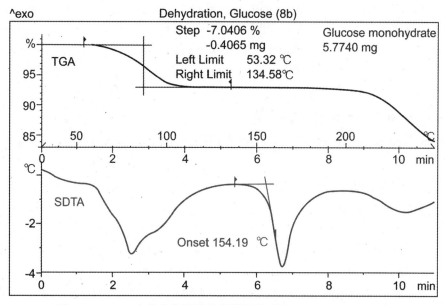

图 3.8 α-D—一水葡萄糖用钻孔盖坩埚测量的 TGA 和 SDTA 曲线
Figure 3.8 TGA and SDTA curves of α-D-Glucose monohydrate measured at a pan with pierced lid

解释 图 3.8 为 α-D—一水葡萄糖的 TGA 和 SDTA 曲线。在盖钻孔的坩埚中的热重测试，与由 DSC 曲线得到的结果解释一致，尤其是由结晶水蒸发产生的失重以及随后的 α-D-无水葡萄糖的熔融。53 ℃ 至 134 ℃ 间 7% 的失重台阶比化学计量值略低，不过，可解释为样品贮存期间结晶水的损失。

Interpretation Figure 3.8 shows the TGA and SDTA cures of α-D-Glucose monohydrate measured at a pan with pierced lid. Thermogravimetric measurements using a pan with a pierced lid confirm the interpretation of the results obtained from the DSC curves, in particular the weight loss caused by the evaporation of the water of crystallization as well as the melting of the α-D-Glucose anhydrate afterwards. The weight loss step of 7% between 53 ℃ and 134 ℃ is somewhat less than that expected stoichiometrically. It can be explained however by a loss of water of crystallization during storage of the sample.

TGA 计算
Evaluation TGA

	温度 Temperature ℃	效应 Effect
TGA 台阶 TGA step	53～134	7%失重(结晶水) 7.0% weight loss (water of crystallization)
SDTA 起始点 SDTA onset	59	吸热峰 endothermic peaks
SDTA 起始点 SDTA onset	154.2	熔融峰 melting peak

结论 含结晶水的物质和它的无水形式通常具有不同的熔点(假多晶型)。如果含结晶水的形式不发生分解，则可在密封坩埚中测定其熔点。在敞口坩埚中，结晶水可逃逸，

Conclusion A substance that contains water of crystallization and its anhydrous form normally have different melting points (pseudopolymorphism). The melting point of the form containing the water of crystallization can be determined in a hermetically sealed pan, provided that no decomposition occurs. In an open pan the

因而测得的是无水形式的熔点。始终应该通过测量失重来确认含结晶水形式的存在。

water of crystallization can escape so that the melting point of the anhydrous form is measured. The presence of a form with water of crystallization should always be confirmed by measuring the weight loss.

3.8 丁基羟基茴香醚的样品制备
Sample Preparation, Butylated Hydroxyanisole

样品 Sample	丁基羟基茴香醚 Butylated Hydroxyanisole		
应用 Application	非活性成分(抗氧剂) Inactive ingredient (antioxidant)		
条件 Conditions	测试仪器：DSC 坩埚：40 μl 铝坩埚，密封。 样品制备： 原样品(1)或在研钵中碾细的晶体。 测试： 以 2.5 K/min 由 30 ℃升温至 70 ℃。 气氛：氮气，50 ml/min	Measuring cell：DSC Pan：Al 40 μl, hermetically sealed. Sample preparation： As received(1) or crystals ground in a mortar(2). Measurement： Heating from 30 ℃ to 70 ℃ at 2.5 K/min. Atmosphere：Nitrogen, 50 ml/min	

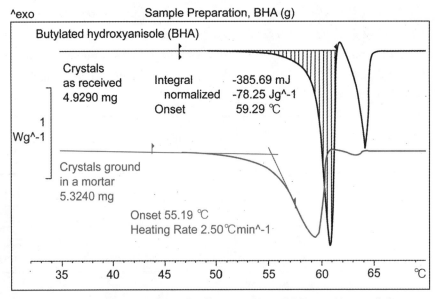

图 3.9 丁基羟基茴香醚原样晶体(上)和碾细后试样(下)的 DSC 曲线

Figure 3.9 DSC curves of Butylated Hydroxyanisole crystals as received (above) and ground (below)

解释 图 3.9 为丁基羟基茴香醚原样品和碾细后试样的 DSC 曲线。两条曲线表明样品制备可能给测量结果带来的影响。两种情况下，可观察到温度范围和熔融热明显不同的两个熔融峰。解释原因为丁基羟基茴香醚的多晶型行为，两个峰对

Interpretation Figure 3.9 displays DSC curves of Butylated Hydroxyanisole crystals as received and ground. The two curves show the effects that sample preparation can have on the results. In both cases, two melting peaks can be observed that differ noticeably in temperature range and in the heats of fusion. The explanation lies in the polymorphic behavior of butylated hydroxyanisole. The two peaks correspond to the possible

应着可能的不同晶型。　　　　crystal modifications.

计算
Evaluation

样品制备 Sample preparation	起始点 1 Onset 1 ℃	ΔH J/g	起始点 2 Onset 2 ℃	ΔH J/g
原样品 as received	59.3	78.2	63.3	27.6
在研钵中碾细 ground in a mortar	55.1	96.8	61.7	1.7

结论 经不同制备的样品(尤其是机械处理)可能导致不同的测量结果。对于呈多晶型的物质尤其如此。

Conclusion A difference in sample preparation (especially mechanical treatment) can lead to different results. This is particularly the case with substances that exhibit polymorphism.

3.9 二丁基羟基甲苯试样量的影响
Influence of the Sample Weight, Butylated Hydroxytoluene

样品 **Sample**	二丁基羟基甲苯 Butylated Hydroxytoluene	
应用 **Application**	非活性成分(抗氧剂) Inactive ingredient (antioxidant)	
条件 **Conditions**	测试仪器：DSC	**Measuring cell**：DSC
	坩埚：40 μl 铝坩埚，密封。	**Pan**：Al 40 μl, hermetically sealed.
	样品制备：原试样。	**Sample preparation**：As received.
	测试：	**Measurement**：
	以 2.5 K/min 由 50 ℃升温至 80 ℃。	Heating from 50 ℃ to 80 ℃ at 2.5 K/min.
	气氛：静态空气	**Atmosphere**：Stationary air

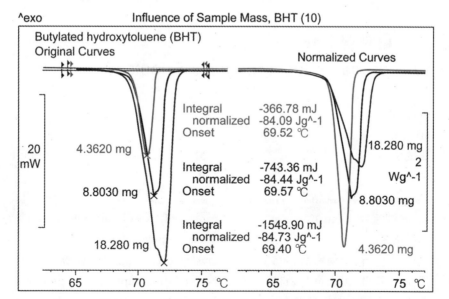

图 3.10　二丁基羟基甲苯不同试样量的 DSC 曲线
Figure 3.10　DSC curves of Butylated Hydroxytoluene

Interpretation Figure 3.10 displays DSC curves of Butylated Hydroxytoluene. The curves show the melting peaks as a function of sample weight. As expected, the peaks in the original presentation (ordinate in mW) increase in height but also in width with increasing weight. Because of this the resolution decreases. In contrast, the normalized presentation in W/g shows that the lowest sample weight gives the highest and narrowest peaks.

Evaluation The onset temperature and heat of fusion of the peaks are determined. The mean values of a number of measurements are presented in the table.

试样量 Sample weight mg	起始点 Onset ℃	熔融焓 ΔH Heat of fusion ΔH J/g
18±0.3	69.4±0.1	85.6, 84.7, 85.6
8.5±0.3	69.6±0.1	83.3, 84.5
4.0±0.4	69.5±0.1	82.6, 84.1, 83.2

Conclusion The sample weight influences the shape of the melting peak. The time required for melting is longer for larger samples because a greater amount of heat has to be transferred. As a result of this, the peaks are shifted to higher temperature. For comparison purposes, the measurement of samples of similar weight is recommended. Samples that are too large are disadvantageous: the peaks become broad (lower resolution) and non-uniform melting leads to irregularly shaped peaks.

3.10 样品贮存和吸湿效应
Sample Storage and Hygroscopic Effects

样品 / Sample: 吸湿性活性成分 / Hygroscopic active ingredient

条件 / Conditions:

Measuring cell: TGA with sample robot

Pan: Al 100 μl, with perforation lid, automatically pierced (needle diameter 1 mm)

Sample preparation: Sample 1: weighed out, pan sealed, lid automatically pierced; Sample 2: weighed out, pan sealed, lid automatically pierced and stored isothermally for 14 hours before measurement at 30 ℃ in the TGA

测试：	**Measurement**：
以 20 K/min 由 30 ℃升温至 150 ℃，所有测试都作了空白修正。	Heating from 30 ℃ to 150 ℃ at 20 K/min, all measurements are blank curve corrected.
气氛：氮气，80 ml/min	**Atmosphere**：Nitrogen, 80 ml/min

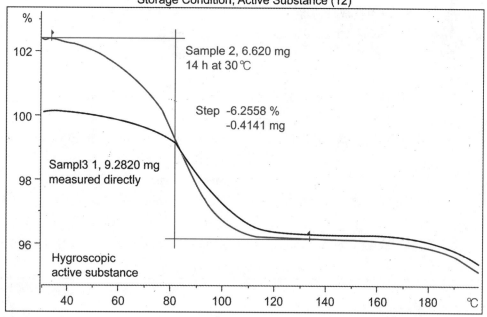

图 3.11 试样 1 和试样 2 的 TGA 曲线

Figure 3.11 TGA curves of Sample 1 and Sample 2

解释 图 3.11 所示分别为样品 1 和样品 2 的 TGA 曲线。该物质非常容易吸湿，对运输、贮存和分析必须采取专门的保护措施。假如对分析本身采取了合适的预防措施，可用 TGA 快速检测水分含量的变化。在本特例中，试样 2 在贮存期间吸收了额外的水分，所以得到不同的测量结果。

Interpretation Figure 3.11 shows the TGA curves of Sample 1 and Sample 2, respectively. The substance is very hygroscopic. Special protective measures must be taken for the transport, storage, processing and analysis of the substance. Changes in the moisture content can be rapidly detected with TGA, provided that suitable precautions are taken for the analysis itself. In this particular case, sample 2 took up additional moisture during the storage time, which is the reason for the different results.

计算
Evaluation

试样序号 Sample No.	台阶 Step %	初始值 Initial value %	经修正的台阶 Corrected step %
1	3.2	100	3.2
2	6.3	102.4	3.3

结论 即使是非常容易吸湿的物质，用密封坩埚，在测试前即钻孔，也可能测定其水分含量。

Conclusion It is possible to measure the water content of even very hygroscopic substances using sealed pans with lids that are pierced immediately before the measurement.

3.11 油的氧化稳定性 Oxidation Stability of Oils

样品	新鲜的和陈旧的玉米油		
Sample	Maize oil, fresh and old		
应用	非活性成分(活性成分的溶剂)		
Application	Inactive ingredient (solvent for active ingredients)		

条件	测试仪器：DSC	**Measuring cell**：DSC
Conditions	坩埚：	**Pan**：
	40 μl 铝坩埚，无盖或 40 μl 铜坩埚，无盖。	Al 40 μl, without lid or copper 40 μl, without lid.
	样品制备：用注射器注满坩埚。	**Sample preparation**：Pan filled with sample using a syringe.
	测试：	**Measurement**：
	以 10 K/min 由 25 ℃ 升温至 300 ℃；125 ℃ 等温。	Heating from 25 ℃ to 300 ℃ at 10 K/min；Isothermal at 125 ℃.
	气氛：氧气或空气,80 ml/min	**Atmosphere**：Oxygen or Air, 80 ml/min

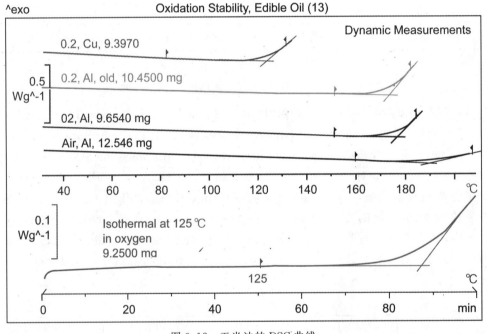

图 3.12　玉米油的 DSC 曲线

Figure 3.12　DSC curves of maize oils

解释　图 3.12 为玉米油的 DSC 曲线。常用动态或等温 DSC 测试来评价油的氧化稳定性,用氧化放热的始点来表征油的氧化稳定性。动态实验的起始点用摄氏度表示；而等温实验则用诱导时间分钟表示。影响起始点的重要因素是坩埚材料(铜具有催化作用)和气氛(氧的反

Interpretation　Figure 3.12 displays DSC curves of the Maize oils. The oxidation stability of oils is often measured with DSC. The measurement is performed dynamically or isothermally. The beginning of the exothermal oxidation is called the onset and characterizes the oil. In a dynamic experiment this is given in degrees Celsius and in an isothermal measurement as the induction time in minutes. Important factors influencing the onset are the material used for the pan (copper acts as a catalyst) and the

应性比空气更强烈)。在相同测试条件下,"陈旧的油"呈现的起始温度要比新鲜的油低,即亦已有部分氧化。

atmosphere (oxygen is more reactive than air). Under the same measurement conditions, the 'old oil' shows an onset temperature lower than that of the fresh oil sample i. e. it is already partially oxidized.

计算 氧化稳定性是由 DSC 曲线氧化起始点表征的。动态实验曲线的轻微漂移,是由油中低分子量成分的开始蒸发引起的,对计算没有影响。在 140 ℃,该失重量约为 0.4%。

Evaluation Oxidation stability is characterized by determining the onset. The slight drift in the curve in the dynamic experiment, which has no effect on the evaluation, is caused by the beginning of evaporation of low molecular weight components in the oil. At 140 ℃ this loss amounts to approximately 0.4%.

条件 Conditions	起始点 Onset ℃	起始点 Onset min
铝坩埚,空气 Aluminum pan, air	188.81	—
铝坩埚,氧气 Aluminum pan, oxygen	176.2	—
铝坩埚,氧气,陈旧的油 Aluminum pan, oxygen, old oil	173.1	—
铜坩埚,氧气 Copper pan, oxygen	122.8	—
铝坩埚,氧气 125 ℃恒温 Aluminum pan, oxygen isothermal at 125 ℃	—	87.5

结论 用 DSC 可研究油的氧化稳定性。必须仔细选择实验条件并保持不变,以获得可重复和可对比的结果。也可用高压 DSC 进行测试,以避免挥发性成分的早期蒸发。

Conclusion The oxidation stability of oils can be investigated with DSC. The experimental conditions must be carefully chosen and kept constant in order to obtain reproducible and comparable results. The measurement can also be performed in a high pressure DSC in order to avoid early evaporation of volatile components.

3.12 香草醛熔融行为的表征
Characterization of the Melting Behavior, Vanillin

样品 香草醛
Sample Vanillin
应用 调味剂
Application Flavoring agent
条件 测试仪器:DSC 和 TGA
Conditions 坩埚:40 μl 铝坩埚,盖钻孔或无盖。
样品制备:原样品。

Measuring cell:DSC and TGA
Pan:Al 40 μl, with pierced lid or open.
Sample preparation:As received.

DSC 测试：	DSC Measurement:
以 10 K/min 由 30 ℃升温至 250 ℃。	Heating from 30 ℃ to 250 ℃ at 10 K/min.
TGA 测试：	TGA Measurement:
以 10 K/min 由 30 ℃升温至 250 ℃，经空白修正。	Heating from 30 ℃ to 250 ℃ at 10 K/min, blank curve corrected
气氛：	Atmosphere:
氮气，DSC：50 ml/min，TGA：20 ml/min	Nitrogen，DSC：50 ml/min，TGA：20 ml/min

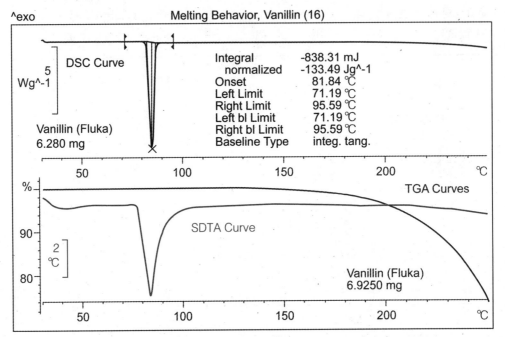

图 3.13 香草醛的 DSC 和 TGA – SDTA 曲线
Figure 3.13 DSC and TGA/SDTA curves of vanillin

解释 图 3.13 为香草醛试样的 DSC、TGA 及 SDTA 曲线。可用热光分析法例如熔点仪由简单的熔融测试得到熔点和熔程。DSC 熔融曲线提供了测试过程中可能会发生的所有效应，例如，曲线在熔融过程后出现基线的轻微漂移。TGA 曲线显示，在该区域没有失重，表明 DSC 基线的漂移是由比热容（c_p）的变化引起的。TGA 的 SDTA 信号可与 TGA 测试一起定性评估熔融行为。

由于香草醛的蒸发和分解，温度高于 150 ℃时失重不断增加。

计算 DSC 峰起始点测定和熔融峰积分。

Interpretation Figure 3.13 displays DSC, TGA and SDTA curves of the Vanillin sample. The melting point and melting range can be gained from simple melting determinations with thermal optical analysis such as a melting poit instrument. However, the DSC melting curve provides a survey of possible events that can occur during the measurement. For instance the curve shows a slight shift of the baseline after the melting process. The TGA curve shows no weight loss in this region, which indicates that the DSC baseline shift is caused by a change in the specific heat capacity (c_p). The SDTA signal of the TGA allows a qualitative evaluation of the melting behavior along with the TGA measurement.

An increasing weight loss occurs at temperatures above 150 ℃ because the vanillin evaporates and decomposes.

Evaluation Onset temperature determination and integration of the melting peak of the DSC curve.

熔点(起始点) Onset ℃	81.8
熔融热 ΔH Heat of fusion ΔH J/g	133.5

结论 DSC 适用于快速测定熔点和熔融热。从 TGA 测试可获得进一步的信息。

Conclusion The DSC is suitable for the rapid determination of melting points and heats of fusion. Additional information can be gained from a TGA measurement.

3.13 胆固醇十四烷酸酯的相转变
Phase Changes, Cholesteryl Myristate

样品 胆固醇十四烷酸酯,细小片
Sample Cholesteryl myristate, fine platelets

条件 测试仪器:DSC
Conditions 坩埚:40 μl 铝坩埚,密封。
样品制备:原样品。
测试:
由 40 ℃升温至 110 ℃;然后降温至 40 ℃。第二次由 40 ℃升温至 110 ℃。
所有升降温速率均为 5 K/min。
气氛:氧气或空气,50 ml/min

Measuring cell: DSC
Pan: Al 40 μl, hermetically sealed.
Sample preparation: As received.
Measurement:
Heating from 40 ℃ to 110 ℃, then cooling down to 40 ℃. Second heating run from 40 ℃ to 110 ℃.
All steps with a heating/cooling rate of 5 K/min.
Atmosphere: Oxygen or Air, 50 ml/min

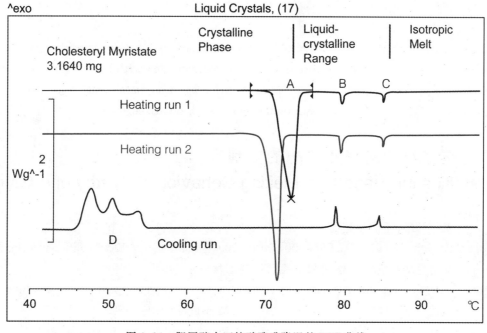

图 3.14 胆固醇十四烷酸酯升降温的 DSC 曲线

Figure 3.14 DSC curves of Cholesteryl myristate measured on heating/cooling

Interpretation Figure 3.14 shows the DSC curves of Cholesteryl myristate measured on heating/cooling Cholesteryl myristate is an example of a substance whose melting process exhibits intermediate liquid crystal stages: On heating, the substance is not directly transformed from the crystalline state to an isotropic melt; the melting process includes a smectic and a cholesteric intermediate stage (mesophases). On cooling at 5 K/min, rapid mesophase transitions and a crystallization process (effect D) can be observed. In the second heating run a slight shift of peak A to lower temperature is noticeable. This is possibly due to better thermal contact between the sample and the pan.

Evaluation Onset temperature determination and integration of the individual peaks.

Peak	A	A	B	B	C	C
Onset and ΔH	起始点 ℃	ΔH J/g	起始点 ℃	ΔH J/g	起始点 ℃	ΔH J/g
First heating run	70.8	67.7	79.2	2.1	84.5	1.7
Cooling			79.1	2.1	84.5	1.7
Second heating run	70.0	66.1	79.2	2.3	84.6	1.8

Conclusion Even mesophase transitions with very low transition enthalpies can be characterized with DSC. Additional information concerning the different mesophases can be gained by the use of thermomicroscopy with polarised light.

3.14 根据熔融行为对聚乙烯醇的鉴别
Identification Based on Melting Behavior, Polyethylene Glycol

Sample Polyethylene glycols, PEG 400, 1000, 2000, 4000, 6000, 10000, different manufacturers

Application Inactive ingredient (basic material for ointments and suppositories, solubilizers etc.)

Conditions Measuring cell: DSC
Pan: Al 40 μl, hermetically sealed.
Sample preparation: As received.

测试:	Measurement:
在 -60 ℃恒温 5min,然后以 10 K/min 升温至 160 ℃。	Held isothermally for 5 minutes at -60 ℃, then heated to 160 ℃ at 10 K/min.
气氛:空气,静止环境,无流动	Atmosphere: Air, stationary environment, no flow rate

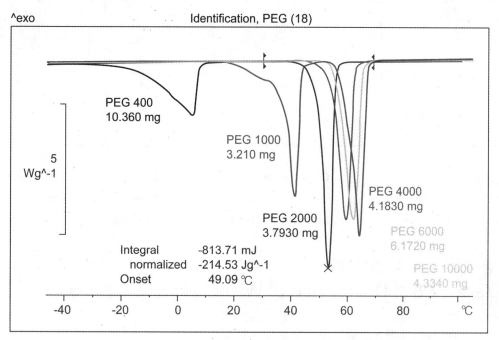

图 3.15　不同链长聚乙烯醇的熔融 DSC 曲线

Figure 3.15　DSC melting curves of polyethylene glycols with different chain lengths

解释　图 3.15 为不同链长聚乙烯醇的熔融 DSC 曲线。

聚乙烯醇是根据平均分子量来命名的。熔点随着链长增加即分子量增大而升高。

链越长,聚乙烯醇熔点之间的差别会越小。这对于清晰区别高分子量样品就更加困难。

物质的纯度越高,熔融峰就越尖锐,对高分子来说,就是微晶尺寸均匀度越高。熔融热变化与链长的关系较小,也与结晶度有关,因而这不是一个旨在区分的合适标准。

计算　测定起始温度和熔融热。

Interpretation　Figure 3.15 shows DSC melting curves of polyethylene glycols with different chain lengths.

Polyethylene glycols are named after their average molecular mass. The melting point increases with increasing chain length, i.e. with increasing molecular mass.

The longer the chain length, the less the melting points of the polyethylene glycols differ from each other. This makes it increasingly difficult to distinguish clearly between samples of high molecular mass.

The melting peak becomes sharper the greater the degree of purity of the substance or, in the case of macromolecules, the greater the degree of uniformity of the size of the crystallites. The differences in the heats of fusion as a function of chain length are relatively small and are also dependent on crystallinity, so that this is not a suitable criterion for differentiation purposes.

Evaluation　Onset temperature determination and the heat of fusion.

样品 Sample	预计熔程(制造商数据) Expected melting range (manufacturer's data) ℃	测量结果 Results	
		起始点 Onset ℃	ΔH J/g
PEG 400	−3～8	−14.3	156.2
PEG 1000	30～40	37.5	194.4
PEG 2000	45～50	49.1	214.5
PEG 4000	50～58	54.8	207.8
PEG 6000	56～63	56.8	212.6
PEG 10000	—	59.4	218.6

结论 可用 DSC 测量熔程来表征聚乙烯醇,即使随着链长增大差别变小。

Conclusion Polyethylene glycols can be characterized by DSC by measuring their melting range, even though the differences become smaller with increasing chain length.

3.15 糖溶液水的熔点降低
Melting Point Depression of Water, Sugar Solutions

样品 **Sample**	葡萄糖、乳糖和蔗糖溶液 Glucose, lactose and saccharose solutions	
应用 **Application**	非活性成分(溶液中稳定剂) Inactive ingredient (stabilizer in solutions)	
条件 **Conditions**	测试仪器:DSC 坩埚:40 μl 铝坩埚,密封。	**Measuring cell**:DSC **Pan**:Al 40 μl, hermetically sealed.
	样品制备: 用小移液管将一滴溶液移入坩埚。	**Sample preparation**: One drop of solution is weighed into the pan using a fine pipette.
	测试: 由 20 ℃降温至−60 ℃,然后溶液在−60 ℃平衡 5min,最后以 5 K/min 升温至 25 ℃。	**Measurement**: Cooling from 20 ℃ to −60 ℃. The solution was then equilibrated for 5 minutes at −60 ℃ and finally heated at 5 K/min to 25 ℃.
	气氛:空气,静止环境,无流动	**Atmosphere**:Air, stationary environment, no flow rate

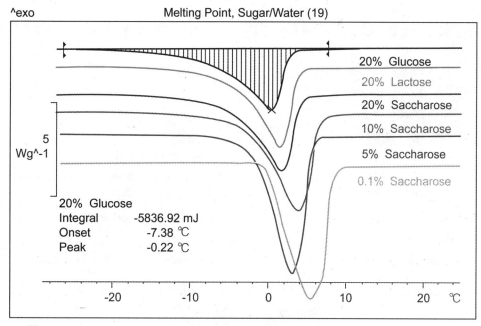

图 3.16　不同糖溶液的 DSC 熔融峰

Figure 3.16　DSC melting peaks of the different sugar solutions

解释　图 3.16 为不同的糖溶液的熔融 DSC 曲线。溶解物的行为如同杂质,溶剂(水)的熔点降低。

Interpretation　Figure 3.16 shows the melting curvess of the different sugar solutions. The dissolved substance behaves as an impurity and depresses the melting point of the solvent (water).

计算　测定溶液的熔融峰温和熔融热。

Evaluation　The onset temperatures of the melting peaks of the solutions and the heats of fusion are determined.

样品 Sample	起始点 Onset ℃	试样量 Sample weight mg
葡萄糖(20%重量) Glucose (20 weight%)	−7.4	20.067
乳糖(20%重量) Lactose (20 weight%)	−4.4	14.666
蔗糖(20%重量) Saccharose (20 weight%)	−4.4	9.412
蔗糖(10%重量) Saccharose (10 weight%)	−2.6	13.221
蔗糖(5%重量) Saccharose (5 weight%)	−1.8	9.715
蔗糖(0.1%重量) Saccharose (0.1 weight%)	−0.3	14.495

结论　加入糖的浓度越大,起始温度越低。

Conclusion　The higher the concentration of added sugar, the lower the onset temperature.

3.16 油包水乳膏的 DSC "指纹"   DSC 'Fingerprint', O/W Cream

样品	乳膏,样品 A 和 B		
Sample	Creams, samples A and B		
应用	非活性成分		
Application	Inactive ingredient		
条件	测试仪器:DSC	**Measuring cell**: DSC
Conditions	坩埚:40 μl 铝坩埚,密封。	**Pan**: Al 40 μl, hermetically sealed.
	样品制备:原样品。	**Sample preparation**: As received.
	测试:以 5 K/min 由 5 ℃升温至 70 ℃。	**Measurement**: Heating from 5 ℃ to 70 ℃ at 5 K/min.
	气氛:氮气,50 ml/min	**Atmosphere**: Nitrogen, 50 ml/min

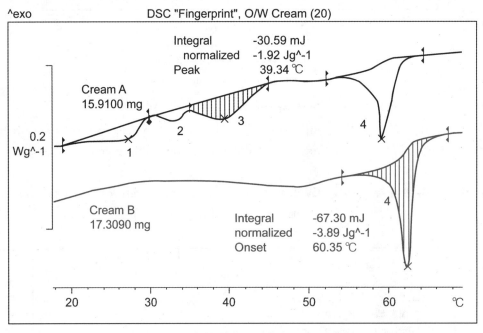

图 3.17 乳膏样品 A 和 B 特有的 DSC 曲线

Figure 3.17 DSC curves characteristic of cream samples A and B

解释 脂肪酸酯和单、二、三甘油酯等这种用于制造油包水乳膏的非活性成分,由于水的加入,可形成不同类型和强度的三维结构。升温时,组分熔融或经历特有的转变,结果得到特定混合物所特有的 DSC 曲线,如图 3.17 所示。

Interpretation The inactive ingredients such as fatty acid esters and mono-, di- and tri-glycerides that are used for the manufacture of O/W creams, can form three dimensional structures of different types and strengths with the addition of water. On heating, the composition melts or undergoes specific transformations. This results in a DSC curve that is characteristic of a particular mixture, as shown in Figure 3.17.

计算 进行了峰温或起始温度(℃)的测定和所有峰的积分(J/g)。根据峰形选用了"样条"或"直线"基线。

Evaluation The peak maxima or onset temperatures (in ℃) and integral ΔH (in J/g) of all the peaks are determined. The baseline types 'spline' or 'line' are used, depending on the shape of the peak.

效应 Effect	1		2		3	
	峰 Peak ℃	ΔH J/g	峰 Peak ℃	ΔH J/g	峰 Peak ℃	ΔH J/g
试样 A Sample A	27.1	1.1	32.5	0.4	58.1	2.8
试样 B Sample B	—	—	—	—	60.4	3.9

结论 特征转变温度和能量可用DSC测量,即使观察到的响应甚低,仍可用作表征油包水乳液的"指纹"。

Conclusion Characteristic transformation temperatures and energies can be measured by DSC. Even if the effects observed are quite weak, they can still be used as a 'fingerprint' to characterize O/W emulsions.

3.17 D,L 丙交酯-乙交酯共聚物的玻璃化转变
Glass Transition, Poly (D, L-lactide)-Co-Glycolide (DLPLGGLU)

样品 Sample	D,L 丙交酯-乙交酯共聚物 Poly(D,L-lactide)-Co-Glycolide	
应用 Application	植入物的生物可降解聚合物 Biologically degradable polymer for implants	
条件 Conditions	测试仪器:DSC	**Measuring cell**:DSC
	坩埚:40 μl 铝坩埚,密封。	**Pan**:Al 40 μl, hermetically sealed.
	样品制备:原样品。	**Sample preparation**:As received.
	测试:	**Measurement**:
	由 30 ℃升温至 160 ℃,然后降温至 −60 ℃并在 −60 ℃恒温 10min,然后升温至 160 ℃,各段的升温或降温速率均为 10 K/min。	Heating from 30 ℃ to 160 ℃, then cooling to −60 ℃, 10 minutes isothermally at −60 ℃, then heating again to 160 ℃, all steps with a heating or cooling rate of 10 K/min.
	气氛:氮气,50 ml/min	**Atmosphere**:Nitrogen, 50 ml/min

解释 图 3.18 为 D,L 丙交酯-乙交酯共聚物的 DSC 曲线。在第 1 次升温期间,呈现试样的热和力学松弛,并将热历史消除。第 2 次升温曲线可清晰观察到玻璃化转变和低能量的松弛峰。

Interpretation Figure 3.18 shows DSC curves of Poly(D,L-lactide)-Co-Glycolide. During the first heating run, thermal and mechanical relaxation occurs and eliminates the thermal history of the sample. In the second heating curve the glass transition and a low energy relaxation peak can be clearly seen.

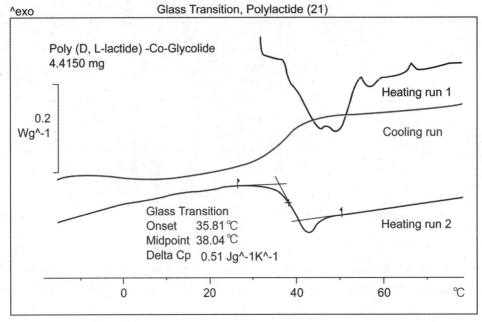

图 3.18　D,L 丙交酯-乙交酯共聚物的 DSC 曲线
Figure 3.18　DSC curves of Poly(D,L-lactide)-Co-Glycolide

计算 Evaluation	玻璃化转变起始点 Onset of Glass transition ℃	玻璃化转变中点 Midpoint of Glass transition ℃	ΔC_p J/gK
	35.8	38.0	0.51

结论　玻璃化转变温度与共聚物中两个组分的比例有关,因此,是一个适合用作质量控制的标准。通常只用第 2 次升温曲线来表征,以保证实际测量前的标准热历史。

Conclusion　The glass transition temperature is a function of the proportions of the two components in the polymer and is therefore a suitable criterion for use in quality control. For characterization purposes, only the second heating curve is normally used in order to ensure a standard thermal history before the actual measurement.

3.18　羟丙基甲基纤维素邻苯二甲酸酯(HPMC-PH)的玻璃化转变和水分含量
Glass Transition and Moisture Content, Hydroxypropoxymethylcellulose Phthalate (HPMC-PH)

样品 Sample	羟丙基甲基纤维素邻苯二甲酸酯(HPMCPH HP 55) Hydroxypropoxymethylcellulose phthalate (HPMCPH HP 55)	
应用 Application	非活性成分(片剂包衣) Inactive ingredient (tablet coating substance)	
条件 Conditions	测试仪器:DSC 和 TGA DSC 坩埚:40 μl 铝坩埚,盖钻孔。 TGA 坩埚:100 μl 铝坩埚,盖钻孔。	**Measuring cell**: DSC and TGA **Pan DSC**: Al 40 μl, with pierced lid. **Pan TGA**: Al 100 μl, with pierced lid.

样品制备：	Sample preparation：
样品在不同相对湿度条件的干燥器中贮存。	The samples were stored under different relative humidity conditions in exsiccators.
DSC 测试：	**DSC Measurement：**
以 20 K/min 由 30 ℃升温至 150 ℃、降温至 30 ℃、在 30 ℃恒温 5min、然后升温至 300 ℃。	Heating from 30 ℃ to 150 ℃, cooling to 30 ℃, 5 minutes isothermally at 30 ℃, then heating to 300 ℃ at 20 K/min.
TGA 测试：	**TGA Measurement：**
以 20 K/min 由 30 ℃升温至 300 ℃。	Heating from 30 ℃ to 300 ℃ at 20 K/min.
气氛：	**Atmosphere：**
氮气，DSC：50 ml/min，TGA：80 ml/min	Nitrogen，DSC：50 ml/min，TGA：80 ml/min

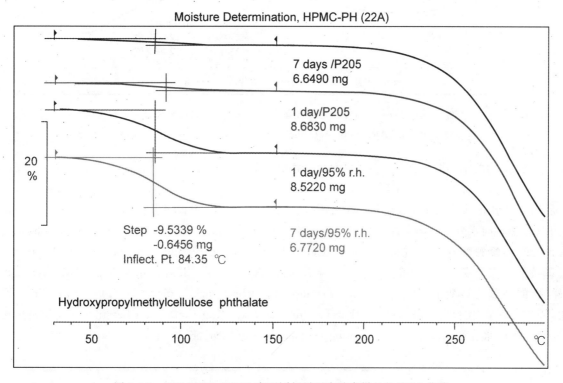

图 3.19　HPMCPH HP 55 在不同相对湿度贮存样品的 TGA 曲线

Figure 3.19　TGA curves of HPMCPH HP 55 samples stored under different relative humidity conditions

解释　图 3.19 为贮存在不同相对湿度条件下样品的 TGA 曲线。试样的水含量可由相应 TGA 曲线的第 1 个台阶测定。得到的值与贮存条件有关。

Interpretation　Figure 3.19 shows the TGA curves of samples that were stored under different relative humidity conditions. The moisture content of a sample can be determined from the first step of the corresponding TGA curve. The values obtained correlate with the storage conditions.

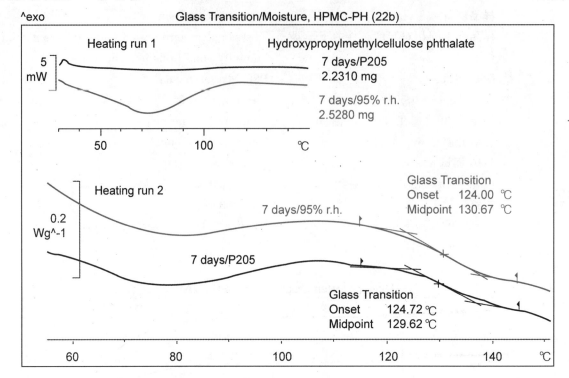

图 3.20 α-D-—水葡萄糖在钻孔盖坩埚中测量的 TGA 和 SDTA 曲线

Figure 3.20　TGA and SDTA cures of α-D-Glucose monohydrate measured at a pan with pierced lid

解释　图 3.20 为两个不同试样第 1 次和第 2 次升温的 DSC 曲线。由于测量是用盖钻孔的坩埚完成的,所以在第 1 次测量时水分可以蒸发掉,可由大的吸热峰观察到这一过程。峰与释放的水量有关。此外,峰与"湿样品"的玻璃化转变部分重叠。因此,第 2 次升温用于测定玻璃化转变(T_g),此时水分业已释放。

Interpretation　Figure 3.20 shows the first and second heating DSC curves of two different samples. Since the measurements were done in a pans with pierced lids, the moisture was able to evaporate in the first heating run. This process can be observed as a large endothermic peak. This correlates with the amount of water released. Apart from this, the peak partially overlaps the glass transition of the 'wet sample'. The glass transition (T_g) is therefore determined using the second heating run, in which the moisture has already escaped.

TGA 计算
Evaluation TGA

贮存条件 Storage	TGA 台阶 TGA Step %	DSC T_g 中点 DSC T_g midpoint ℃
P_2O_5 环境 7 天 7 days over P_2O_5	1.18	129.6
P_2O_5 环境 1 天 1 days over P_2O_5	1.39	—
未处理 Untreated	2.50	—
相对湿度 95% 1 天 1 day at 95% r. h.	8.28	—
相对湿度 95% 7 天 7 days at 95% r. h.	9.50	130.7

结论 纤维素类聚合物易于吸水，吸水量取决于贮存条件。通常用无水样品测定玻璃化转变温度（T_g），由第 2 次升温曲线测得。如果必须对含不同水分含量的聚合物测定玻璃化转变温度，则必须使用密封坩埚。

Conclusion Cellulose-based polymers take up moisture easily. The uptake is dependant on the storage conditions. The moisture-free sample is normally used for determining the glass transition temperature (T_g). This is done by evaluating the second heating run. If the glass transition temperature has to be determined with different moisture contents, then sealed pans have to be used.

3.19 聚乙烯薄膜的质量控制　Quality Control, PE Films

样品	线型低密度聚乙烯(PE-LLD)、低密度聚乙烯（PE-LD）、高密度聚乙烯（PE-HD）	
Sample	PE-LLD, PE-LD, PE-HD	
应用	包装材料	
Application	Packaging material	
条件	测试仪器：DSC	**Measuring cell**：DSC
Conditions	坩埚：40 μl 铝坩埚，密封。	**Pan**：Al 40 μl, hermetically sealed.
	样品制备：由薄膜切割出的试样。	**Sample preparation**：Samples cut out of the films.
	测试：	**Measurement**：
	由 30 ℃升温至 160 ℃，降温至 30 ℃，然后升温至 160 ℃。各段的升温或降温速率均为 10 K/min	Heating from 30 ℃ to 160 ℃, cooling to 30 ℃, then heating to 160 ℃, all steps with heating or cooling rates of 10 K/min.
	气氛：氮气，50 ml/min	**Atmosphere**：Nitrogen, 50 ml/min

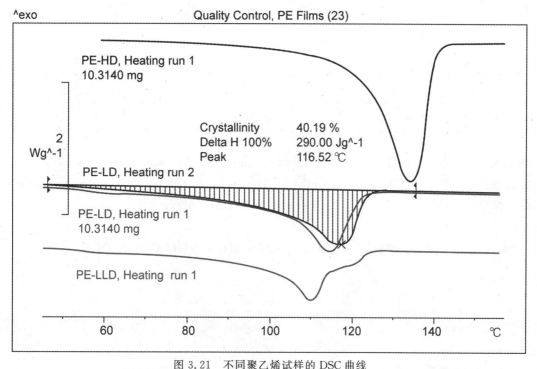

图 3.21　不同聚乙烯试样的 DSC 曲线

Figure 3.21　DSC curves of different PE samples

解释　图 3.21 为不同聚乙烯试样的 DSC 曲线。

Interpretation　Figure 3.21 shows the DSC curves of different PE samples.

DSC 是美国药典中指定的聚乙烯容器物理测试方法：包装材料的质量对于原料和最终产品的保证至关重要。尤其是基本包装材料，即直接与产品接触的包装材料，有关质量必须完美无缺，必须满足指标要求，绝不可与需要保护的物质发生作用。第 1 次升温的作用是消除任何（未知的）热历史，因而不同质量的材料能在第 2 次升温进行合适的比较。峰温和结晶度（由熔融热计算得到）用作质量标准。

DSC is specified in the USP for the physical testing of polyethylene containers: the quality of packaging material is of decisive importance for the protection of raw materials and end products. In particular, primary packaging material i.e. packaging material that comes into direct contact with the product must be perfect as far as quality is concerned, must satisfy the specified requirements and must not react with the material that it is designed to protect. The first heating run serves to eliminate any (unknown) thermal history so that samples of different quality can be properly compared with each other in the second heating run. The melting peak temperature and the crystallinity (calculated from the heat of fusion), are used as quality criteria.

计算 样品的结晶度由测量得到的熔融热与 100% 结晶的 PE 的熔融热（设定其熔融热为 290 J/g）之比测定，由图 3.21 中曲线之一作为例子做了说明。

Evaluation The crystallinity of a sample is determined by comparing its measured heat of fusion with that of 100% crystalline PE which is assumed to have a heat of fusion of 290 J/g. This is demonstrated as example with one of the curves in figure 3.21.

	第 1 次升温 Heating run 1		第 2 次升温 Heating run 2	
试样 Sampe	峰 Peak ℃	结晶度 Crystallinity %	峰 Peak ℃	结晶度 Crystallinity %
PE-LD	113.7	37.9	116.5	40.2
PE-HD	132.9	64.5	—	—
PE-LLD	109.6	31.7	109.8	31.4

结论 用 DSC，很容易地可将样品相互区分开来。该方法非常适于旨在快速的质量控制。

Conclusion Using DSC, the samples can be distinguished from each other without any difficulty. The method is very suitable for rapid quality control purposes.

3.20 氢化可的松的分解　Decomposition, Hydrocortisone

样品 Sample	氢化可的松 Hydrocortisone
应用 Application	活性成分 Active ingredient
条件 Conditions	测试仪器：TGA　　　　　　Measuring cell：TGA 坩埚：70 μl 氧化铝坩埚　　　Pan：Alumina 70 μl 样品制备：原样品　　　　　Sample preparation：As received

测试：	**Measurement**：
以 20 K/min 由 25 ℃升温至 800 ℃，作空白曲线修正。	Heating from 25 ℃ to 800 ℃ at 20 K/min, blank curve corrected.
气氛：空气，200 ml/min	Atmosphere：Air，200 ml/min

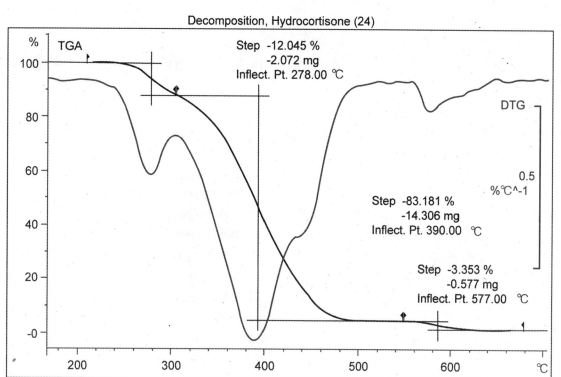

图 3.22　氢化可的松的 TGA 和 DTG 曲线

Figure 3.22　TGA and DTG curves of Hydrocortisone

解释　可用 TGA 测量该物质以若干个台阶发生的分解，如图 3.22 所示。一阶微商 DTG 曲线有助于清晰分辨使各个台阶。

Interpretation　The decomposition of this substance, which takes place in several steps can be measured with TGA, as shown in figure 3.22. The derivative curve, DTG curve, is helpful in making the steps more evident.

计算　失重用水平切线表示于 TGA 曲线上。测定所用的界限是用 DTG 曲线确定的。

Evaluation　Weight loss is shown on the TGA curve with horizontal tangents. The limits used for the determination are set by making use of the DTG curve.

台阶 Step	台阶高度 Step height %	计算范围 Evaluation range ℃
台阶 1 Step 1	12.0	175~285
台阶 2 Step 2	83.2	285~538
台阶 3 Step 3	3.4	538~690

结论 用 TGA 可定量测定各个失重。不过,要确定反应类型或逸出物质的性质,必须使用其他分析方法。

Conclusion With TGA, the individual weight losses can be quantified. Additional analytical methods must be used, however, in order to determine the type of reaction or the nature of the evolved substances.

3.21 甲磺酸双氢麦角胺熔点处的分解
Decomposition at the Melting Point, Dihydroergotamine Mesylate

样品 **Sample**	甲磺酸双氢麦角胺 Dihydroergotamine mesylate		
应用 **Application**	活性成分 Active ingredien		
条件 **Conditions**	测试仪器:DSC 和 TGA 坩埚:40 μl 或 100 μl 铝坩埚,盖钻孔。 样品制备:原样品。 DSC 测试: 以 20 K/min 由 30 ℃升温至 300 ℃。 TGA 测试: 以 20 K/min 由 30 ℃升温至 300 ℃,经空白修正。 气氛: 氮气,DSC:50 ml/min,TGA:80 ml/min	**Measuring cell**:DSC and TGA **Pan**:Al 40 μl or 100 μl, with pierced lid. **Sample preparation**:As received. **DSC Measurement**: Heating from 30 ℃ to 300 ℃ at 20 K/min. **TGA Measurement**: Heating from 30 ℃ to 300 ℃ at 20 K/min, blank curve corrected. **Atmosphere**: Nitrogen,DSC:50 ml/min,TGA:80 ml/min	

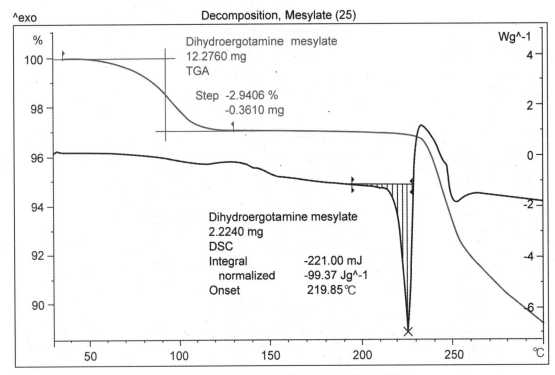

图 3.23 甲磺酸双氢麦角胺的 DSC 和 TGA 曲线

Figure 3.23 DSC and TGA curves of Dihydroergotamine mesylate

解释　图 3.23 为甲磺酸双氢麦角胺的 DSC 和 TGA 曲线。该物质在 219 ℃熔点处分解，呈现为 DSC 曲线上的放热峰。这可由 TGA 曲线在此区域呈现剧烈失重确认这种解释。在此之前更低温度，试样失去含量约为 3% 的水分，这在 DSC 曲线上仅可观察到轻微的迹象。

由于熔融与分解过程重叠，所以应该仔细处理熔融热的测量。选择较快的升温速率可将分解移至较高温度，改善数据计算。

Interpretation　Figure 3.23 shows the DSC and TGA curves of Dihydroergotamine mesylate. The substance decomposes at its melting point of 219 ℃. An indication of this is the exothermic peak in the DSC curve. This interpretation is confirmed by the TGA curve which shows a drastic loss of weight in this region. Before this, at lower temperatures, the sample loses moisture, the calculated content being approximately 3%. Only slight indications of this can be seen in the DSC curve.

The measurement of the heat of fusion should be treated with care since the melting and decomposition processes overlap. The selection of a faster heating rate could possibly shift the decomposition to higher temperature and allow a better evaluation.

计算
Evaluation

DSC 熔融峰 DSC melting peak		TGA 干燥台阶 TGA drying step	
起始点 Onset ℃	熔融热 ΔH Heat of fusion ΔH J/g	台阶 Step %	计算范围 Evaluation range ℃
219.9	99.4	2.9	35—130

结论　通常，仅用一种分析技术不可能完全解释热效应。只有联合使用不同方法或改变测试条件，才有可能获得发生过程的较全面信息。如遇伴有分解的物质，则只能用快速升温速率测定其熔点。经典熔点仪无法提供所要求的 20 K/min～50 K/min 的升温速率。

Conclusion　Frequently, it is not possible to explain thermal effects completely with just one analytical technique. Only when a combination of different methods or variations of measurement conditions are employed is it possible to gain a more complete picture of the processes taking place. Melting points of substances that decompose can only be determined using high heating rates. The required heating rates of 20～50 K/min can not be attained with classical melting point equipment.

3.22　阿斯巴甜的熔融和分解
Melting Behavior and Decomposition, Aspartame

样品　阿斯巴甜
Sample　Aspartame

应用　非活性成分（甜味剂）
Application　Inactive ingredient (sweetening agent)

条件　测试仪器：DSC 和 TGA
Conditions　
坩埚：100 μl 铝坩埚，盖钻孔。
样品制备：原样品。
DSC 测试：
以 10 K/min 由 40 ℃升温至 300 ℃。

Measuring cell：DSC and TGA
Pan：Al 100 μl, with pierced lid.
Sample preparation：As received.
DSC Measurement：
Heating from 40 ℃ to 300 ℃ at 10 K/min.

TGA 测试:

以 20 K/min 由 30 ℃升温至 300 ℃。

气氛:氮气,80 ml/min

TGA Measurement:

Heating from 30 ℃ to 300 ℃ at 10 K/min.

Atmosphere: Nitrogen, 80 ml/min

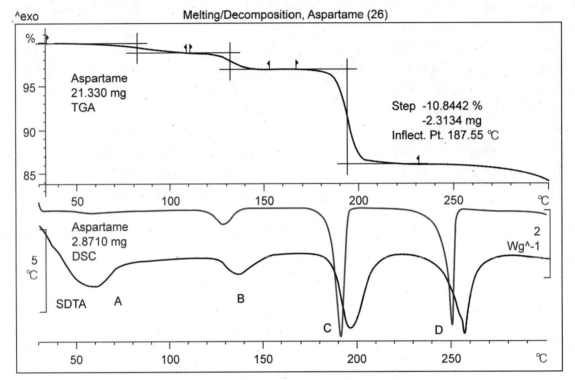

图 3.24 阿斯巴甜的 TGA 和 DSC 及 SDTA 曲线

Figure 3.24 TGA and DSC as well as SDTA curves of aspartame

解释 图 3.24 为阿斯巴甜的 TGA（上）和 DSC（中）及 SDTA（下）曲线。在 TGA 曲线上,该物质呈现若干个失重台阶。同步差热 SDTA 信号表明,所有反应都是吸热的。DSC 曲线也给出了相同的信息,但峰比 SDTA 信号更尖锐,因为其时间常数较小。

在 40 ℃至 110 ℃范围,可观察到缓慢而连续的由失水产生的失重。在约 129 ℃的台阶为失去结晶水,可用质谱仪同步测量来确认。最后,约 187 ℃,该物质分解逸出甲醇（m/e=31）。DSC 和 SDTA 曲线上 245.7 ℃观察到的峰是分解产物（3-羧甲基-6-苯甲基-2,5-二酮哌嗪）的熔点。

Interpretation Figure 3.24 shows TGA (upper) and DSC (middle) as well as SDTA (bottom) curves of aspartame. In the TGA curve the substance shows several weight loss steps. The SDTA signal recorded at the same time shows that all the reactions are endothermic. The DSC curve also gives the same information. The peaks are however sharper than in the SDTA signal because of the shorter time constant.

A slow but continuous weight reduction caused by the loss of moisture can be observed in the region between 40 ℃ and 110 ℃. The step at about 129 ℃ is the elimination of the water of crystallization. This is confirmed by simultaneous measurements with the mass spectrometer. Finally, at about 187 ℃, the substance decomposes by elimination of methanol (m/e = 31). The peak that can be observed at 245.7 ℃ in the DSC and SDTA curves corresponds to the melting of the decomposition product (3-carboxymethyl-6-benzyl-2,5-dioxopiperazine).

计算
Evaluation

峰 Peak	SDTA 起始点 SDTA Onset ℃	DSC 起始点 DSC Onset ℃	DSC ΔH J/g	效应 Effect
A	40.0	—	—	水分 moisture
B	122.3	122.7	29.8	结晶水 water of crystallization
C	183.6	185.4	141.4	分解 decomposition
D	239.4	245.7	121.3	分解产物的熔点 melting of the decomposition product

台阶 Step	TGA 计算范围 TGA Eval. Range ℃	TGA 起始点 TGA Onset ℃	TGA 台阶 TGA Step %	效应 Effect
A	30—110	79.6	1.1	水分 moisture
B	110—150	123.6	1.9	结晶水 water of crystallization
C	180—225	184.1	10.8	分解 decomposition

结论 TGA 和 DSC 测试结果的结合常常有助于对复杂曲线的正确解释。

Conclusion The combination of TGA and DSC results is often useful for the correct interpretation of complicated curves.

3.23 丙二酸的完全分解   Total Decomposition, Malonic Acid

样品 **Sample**	丙二酸 Malonic acid	
应用 **Application**	非活性成分 Inactive ingredien	
条件 **Conditions**	测试仪器：DSC 和 TGA	Measuring cell：DSC and TGA
	坩埚：40 μl 或 100 μl 铝坩埚,盖钻孔。	Pan：Al 40 μl or 100 μl, with pierced lid.
	样品制备：原样品。	Sample preparation：As received.
	DSC 测试：	DSC Measurement：
	以 20 K/min 由 30 ℃升温至 300 ℃。	Heating from 30 ℃ to 300 ℃ at 20 K/min.
	TGA 测试：	TGA Measurement：
	以 20 K/min 由 30 ℃升温至 300 ℃, 经空白修正。	Heating from 30 ℃ to 300 ℃ at 20 K/min, blank curve corrected
	气氛：	Atmosphere：
	氮气,DSC：50 ml/min,TGA：80 ml/min	Nitrogen, DSC：50 ml/min, TGA：80 ml/min

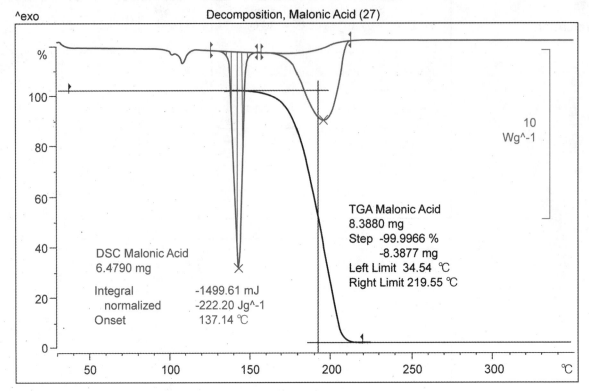

图 3.25 丙二酸的 DSC 和 TGA 曲线

Figure 3.25 DSC and TGA curves of malonic acid

解释 图 3.25 为丙二酸的 DSC 和 TGA 曲线。在约 137 ℃ 熔融后,该物质在吸热过程中分解为乙酸,放出 CO_2,可在 TGA 曲线上观察到 100% 的失重以及在 DSC 曲线上观察到分解吸热峰。基线的移动是由于质量改变,而不是由比热容(c_p)变化产生的。

Interpretation Figure 3.25 shows the DSC and TGA curves of Malonic acid. After melting at about 137 ℃, the substance decomposes to acetic acid vapor in an endothermic process with the liberation of CO_2. This can be seen in the TGA curve as a 100 per cent weight loss as well as in the DSC curve as an endothermic decomposition peak. The displacement of the baseline is a result of the change of mass and is not caused by a change in the specific heat capacity (c_p).

计算
Evaluation

效应 Effect	DSC 起始点 DSC Onset ℃	DSC ΔH J/g	TGA 台阶 TGA Step %	TGA 计算范围 TGA Eval. Range ℃
转变 Transformation	103.8	20.0	—	—
熔融 Melting	137.1	222.2	—	—
分解 Decomposition	174.9	281.5	99.99	35—220

结论 分解可用 DSC 和 TGA 通过热变化或失重来检测。分解产物

Conclusion Decomposition can be detected with DSC and TGA by a heat change or a loss of weight. Further investigation of the

的进一步研究需要用联用技术,例如 TGA/MS 或 TGA/FTIR。

decomposition products requires the use of combination techniques such as TGA-MS or TGA-FTIR.

3.24 乙酰水杨酸分解的动力学分析
Kinetic Analysis of Decomposition, Acetylsalicylic Acid

样品 **Sample**	乙酰水杨酸 Acetylsalicylic acid		
应用 **Application**	止痛剂 Analgesic		
条件 **Conditions**	测试仪器:TGA 坩埚:100 μl 铝坩埚 样品制备:原样品 测试: 以 1 K/min、2 K/min、5 K/min 和 10 K/min 由 25 ℃升温至 300 ℃,所有测试作空白曲线修正。 气氛:空气,50 ml/min	**Measuring cell**:TGA **Pan**:Al 100 μl **Sample preparation**:As received **Measurement**: Heating from 25 ℃ to 300 ℃ at 1, 2, 5 and 10 K/min, all measurements are blank curve corrected. **Atmosphere**:Air, 50 ml/min	

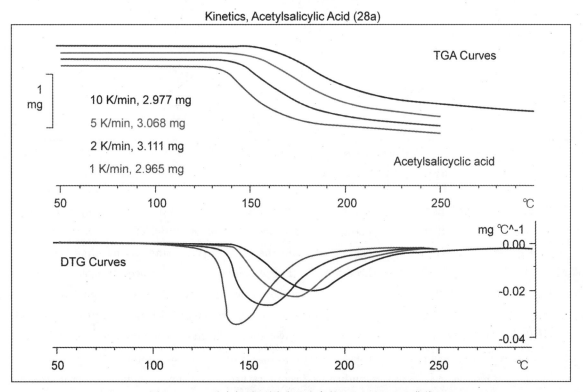

图 3.26 乙酰水杨酸不同升温速率的 TGA 和 DTG 曲线
Figure 3.26 TGA and DTG curves of acetylsalicylic acid at different heating rates

解释 图 3.26 表示升温速率对乙酰水杨酸分解的影响。随着升温速率的提高,分解失重移至较高温度。

Interpretation Figure 3.26 shows the effect of the heating rate on the decomposition of acetylsalicylic acid. The decomposition, that is the resulting weight loss, is shifted to higher temperature

为计算百分转化率曲线,形成 TGA 曲线一阶微商的 DTG 曲线。以便作进一步的动力学分析。选用合适的基线,DTG 曲线就能将第 1 个分解台阶与进一步的重叠分解反应分开(TGA 曲线并未呈水平结束)。

with increasing heating rates. The DTG curves, i. e. the first derivative of the TGA curves, are formed in order to calculate the percentage conversion curves that are used for further kinetic analysis. The DTG curves together with the choice of a suitable baseline allow the separation of the first decomposition step from further overlapping decomposition reactions (the TGA curves do not terminate horizontally).

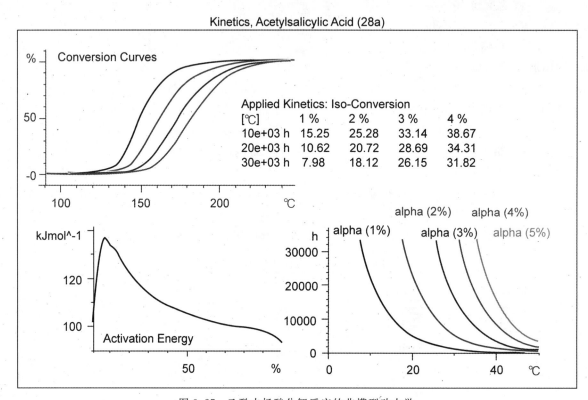

图 3.27 乙酰水杨酸分解反应的非模型动力学

Figure 3.27　Model-free kinetics of acetylsalicylic acid decomposition reaction

解释　图 3.27 所示为用非模型动力学所作的完整数据处理。转化率曲线用"积分水平"基线由 DTG 曲线计算得到。这些转化率曲线转而为非模型动力学计算活化能的基础。活化能介于 100 kJ/mol～140 kJ/mol 范围,与转化率有关,表明是一个复杂反应。活化能可对其他条件(温度和时间)下的分解反应进行模拟(预测),图中用曲线和表格形式表示,称为等转化率曲线和数表。如此表示,则一定温度下物质达到一定百分转化率(或换言之为分解程度)所需要的时间就一目了然。

Interpretation　Figure 3.27 shows the complete evaluation using the model-free kinetics method. The conversion curves are calculated from the DTG curves using the baseline type 'integral horizontal'. These conversion curves are, in turn, the basis for the model-free kinetics for the calculation of the activation energy. This lies in the range 100 kJ/mol～140 kJ/mol and is a function of conversion, indicating a complex reaction. The activation energy allows a simulation of the decomposition reaction for other conditions (temperature and time). This is shown graphically and in tabular form in so-called iso-conversion plot and tables. In this presentation, the time required for the substance to reach a given percentage conversion (or in other words degree of decomposition) at a given temperature is evident.

计算 确定活化能与百分转化率的关系,细节见图3.26和3.27。

对转化率1%、2%、3%或4%的贮存时间/贮存温度进行预测(应用动力学)。由表可知,当该物质在15.3 ℃贮存了10000 h后,转化率达到1%;1年有8760 h。

Evaluation The activation energy is determined as a function of the percentage conversion. The details are given in figure 3.26 and figure 3.27.

Prediction of the storage time/storage temperature for a conversion of 1, 2, 3 or 4% (applied kinetics). It is evident from the table, that a conversion of 1% is reached when the substance is stored at 15.3 ℃ for 10 000 hours; 1 year has 8760 hours.

时间 Time	转化率 Conversion			
	1%	2%	3%	4%
10000h	15.3 ℃	25.3 ℃	33.1 ℃	38.7 ℃
20000h	10.6 ℃	20.7 ℃	28.7 ℃	34.3 ℃
30000h	8.0 ℃	18.1 ℃	26.1 ℃	31.8 ℃

备注:由液态反应的结果外推至固态反应,会有较高的不确定性。
Please note: Extrapolation of results from reactions performed in the liquid state to the solid state have a high degree of uncertainty.

结论 本例表明,应用热分析非模型动力学是估算药物制剂潜在贮存周期最廉价的一种方法。当然,只有分解与显著失重相一致时才可使用TGA。

必须强调,该方法并不可取代合适的长期测试。该方法对于配方预选更为有用。然后,可对良好性能的配方做费时昂贵的最终测试。

Conclusion The example shows that thermal analysis and the application of model-free kinetics is an efficient means of estimating the potential storage lifetime of pharmaceutical preparations at a minimum expense. The TGA can of course only be used when the decomposition is coincidental with a significant weight loss.

It must be emphasized that this procedure can never replace a proper long-term test. The method is more useful for the preliminary selection of formulations. The formulations that exhibit good properties can then be subjected to the time-consuming and expensive final tests.

3.25 茶碱的水合稳定性 Hydrate Stability, Theophylline

样品 不同的一水茶碱
无水茶碱
Sample Theophylline monohydrate, various samples
Theophylline anhydrous form

应用 活性成分
Application Active ingredient

条件
Conditions
测试仪器:TGA
坩埚:100 μl铝坩埚,盖钻孔
样品制备:原样品
测试:
以20 K/min由30 ℃升温至300 ℃,作空白曲线修正。

Measuring cell:TGA
Pan:Al 100 μl, with pierced lid
Sample preparation:As received
Measurement:
Heating from 30 ℃ to 300 ℃ at 20 K/min, blank curve corrected.

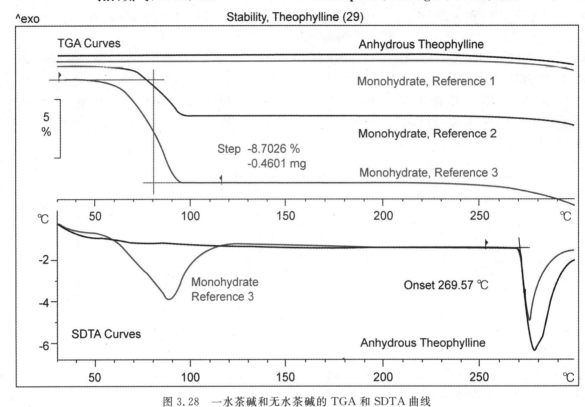

图 3.28　一水茶碱和无水茶碱的 TGA 和 SDTA 曲线

Figure 3.28　TGA and SDTA curves of samples of theophylline monohydrate and anhydrous form

解释　图 3.28 为几种一水茶碱和无水茶碱试样的 TGA 和 SDTA 曲线。

一水茶碱在正常湿度条件下并不稳定,如果没有专门的保护,在贮存过程会失去结晶水。残余水在 TGA 曲线呈现为一个台阶。可见在贮存期间,参比样品 2 已经失去几乎一半的结晶水,参比样品 1 已经完全失去结晶水。因此,该物质贮存时,应该放满容器,完全密封。由于这种参比物质用作标准物或其他方法的参比样品,所以在进行任何计算之前,测定活性物质的实际含量很重要。

同步差热 SDTA 曲线计算表明,一水茶碱(参比样品 3)吸热失去结晶水,以及无水茶碱熔融,熔点约为 269.6 ℃。

计算　由 TGA 曲线测定失重,用

Interpretation　Figure 3.28 shows the TGA and SDTA curves of samples of theophylline monohydrate and theophylline anhydrous form.

Storage of the theophylline monohydrate without special protection leads to loss of the water of crystallization because it is unstable under normal conditions of humidity. The TGA curves show this residual water as a step. It can be seen that during storage reference sample 2 has lost almost half of its water of crystallization and that reference sample 1 has lost it completely. The substance should therefore be stored, tightly sealed, in a well-filled container. Since such reference substances are used as standards or reference samples for other methods, it is important to determine the actual content of the active substance before performing any calculations.

The evaluation of the SDTA curves shows the endothermic loss of water of crystallization from the monohydrate (reference sample 3) as well as the melting of the anhydrous form with the melting point at about 269.6 ℃.

Evaluation　The weight loss is determined from the TGA

水平台阶进行计算。一水茶碱的化学计量预计值为9.1%。

curve. The evaluation is performed using horizontal steps. The value expected stoichiometrically for the monohydrate is 9.1%.

	台阶 Step %	计算范围 Eval. range ℃	试样量 Sample weight mg
一水茶碱参比样品1(陈旧的) Ref. 1 Monohydrate (old)	<0.1	32—120	11.988
一水茶碱参比样品2(陈旧的) Ref. 2 Monohydrate (old)	4.2	32—120	11.452
一水茶碱参比样品1(新鲜的) Ref. 3 Monohydrate (new)	8.7	32—120	5.287
无水茶碱 Anhydrous form	<0.1	32—120	10.01

结论 热重分析是快速定量测定结晶水的方法,并只需少量样品。

Conclusion Thermogravimetric analysis is a method of the rapid quantification of water of crystallization, especially since the technique requires only small amounts of sample.

3.26 淀粉/羟甲基纤维素钠(羧甲基淀粉钠)的水分
Moisture, Starch/NaCMC (Primojel)

样品 Sample	羟甲基纤维素钠/马铃薯淀粉(羧甲基淀粉钠)混合物,水分含量约6% Mixture of NaCMC/potato starch (Primojel), water content approx. 6%	
应用 Application	羧甲基淀粉钠由羟甲基纤维素钠和马铃薯淀粉组成,在水中可吸收超过自身重量40%的水。吸水能力使它可用作片剂的崩解剂。 Primojel consists of Na-CMC and potato starch and is able to take up over 40% of its weight in water. This capacity to take up water leads to its use as a disintegrating agent for tablets.	
条件 Conditions	测试仪器:TGA 坩埚:100 μl 铝坩埚,盖钻孔 样品制备: 原样品,在不同条件下贮存 测试: 以 20 K/min 由 30 ℃升温至 300 ℃,作空白曲线修正。 气氛:氮气,80 ml/min	Measuring cell: TGA Pan: Al 100 μl, with pierced lid Sample preparation: As received, stored under different conditions Measurement: Heating from 30 ℃ to 300 ℃ at 20 K/min, blank curve corrected. Atmosphere: Nitrogen, 80 ml/min

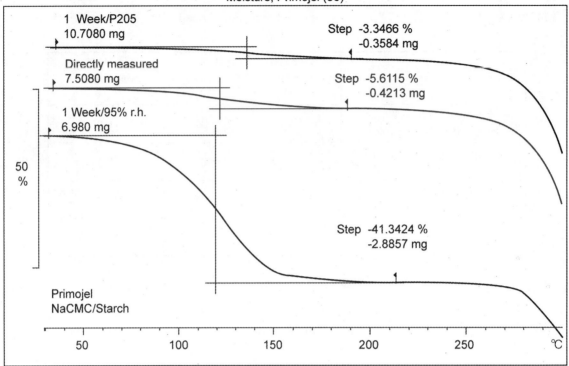

图 3.29 几种羧甲基淀粉钠混合物的 TGA 曲线
Figure 3.29 TGA curves of samples of the different Primojel mixtures

解释 图 3.29 为几种羟甲基纤维素钠/马铃薯淀粉（羧甲基淀粉钠）混合物的 TGA 曲线。试样在约 120 ℃失去水分，失水程度取决于实际贮存条件。相对湿度越高，贮存时间越长，则吸收的水量越大。

Interpretation Figure 3.29 shows the TGA curves of samples of the different mixtures of NaCMC/potato starch (Primojel). The samples lose their moisture at about 120 ℃. The extent of the moisture-loss depends on the actual storage conditions. The higher the relative humidity and the longer the storage time, the greater the amount of water absorbed.

计算
Evaluation

贮存条件 Storage conditions	台阶 Step %
P_2O_5 环境 7 天 7 days over P_2O_5	3.4
P_2O_5 环境 1 天 1 days over P_2O_5	3.4
原样品 As received	5.6
95% 相对湿度 1 天 1 day at 95% rel. humidity	24.8
95% 相对湿度 7 天 7 day at 95% rel. humidity	41.3

结论 当表征对水敏感的样品时，必须考虑贮存条件的影响。

Conclusion The influence of the storage conditions must be taken into account when characterizing samples that are sensitive to moisture.

3.27 三棕榈精的多晶型 Polymorphism, Tripalmitin

样品	三棕榈精		
Sample	Tripalmitin		
应用	非活性成分		
Application	Inactive ingredient		
条件	测试仪器：DSC	**Measuring cell**：DSC	
Conditions	坩埚：40 μl 铝坩埚，密封。	**Pan**：Al 40 μl, hermetically sealed.	
	样品制备：原样品。	**Sample preparation**：As received.	
	测试：	**Measurement**：	
	由 30 ℃升温至 90 ℃，降温至 30 ℃。第 2 次升温由 30 ℃至 90 ℃。各段的升温或降温速率均为 5 K/min	Heating from 30 ℃ to 90 ℃. Cooling to 30 ℃. Second heating run from 30 ℃ to 90 ℃. All steps were performed with heating or cooling rates of 5 K/min.	
	气氛：氮气，80 ml/min	**Atmosphere**：Nitrogen，80 ml/min	

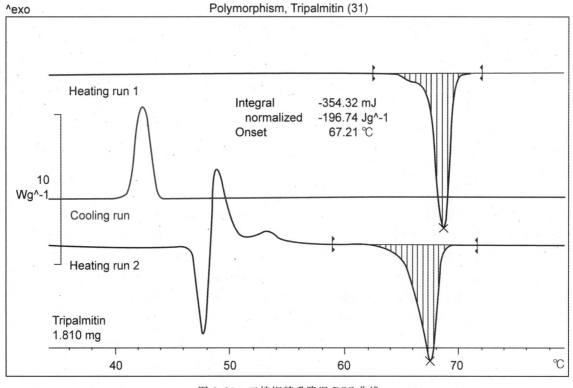

图 3.30 三棕榈精升降温 DSC 曲线

Figure 3.30 DSC curves of tripalmitin at heating runs and cooling

解释 图 3.30 为三棕榈精在第 1 次、第 2 次升温和降温时的 DSC 曲线。第 1 次升温表明，三棕榈精原样品为稳定型的。冷却时，结晶析出亚稳态晶型，其熔点低于热力学稳定态的晶型。第 2 次升温时，熔

Interpretation Figure 3.30 shows DSC curves of tripalmitin at the first and second heating runs and the cooling run. The first heating run shows that tripalmitin as received is in the stable form. On cooling, a metastable modification crystallizes out, whose melting point lies lower than that of the thermodynamically stable form. In the second heating run, the liquid phase formed

融形成的液相在稳定晶型的晶核上结晶。亚稳态晶型熔融热的测定是不确定的(放在括号内),因为在亚稳态晶型熔融的同时发生稳定态晶型的结晶。较低的起始温度和略低的熔融热是由稳定态晶型的不完全结晶造成的。

on melting crystallizes on nuclei of the stable modification. A determination of the heat of fusion of the metastable form is uncertain (and is put in brackets), since the crystallization of the stable form occurs at the same time as the melting of the metastable form. The lower onset temperature and the slightly lower heat of fusion are caused by incomplete crystallization of the stable form.

计算
Evaluation

	起始点 Onset ℃	ΔH J/g	起始点 Onset ℃	ΔH J/g
第1次升温 Heating run 1	—	—	67.2	196.7
降温 Cooling run	43.8	113.9	—	—
第2次升温 Heating run 2	46.7	(66.8)	65.4	190.6

结论 DSC 是检测多晶型极好的技术。

Conclusion The DSC is an excellent technique for the detection of polymorphic modifications.

3.28　甲苯磺丁脲的多晶型　Polymorphism, Tolbutamide

样品 **Sample**	甲苯磺丁脲 Tolbutamide	
应用 **Application**	活性成分(口服抗糖尿病药) Active ingredient (oral antidiabetic)	
条件 **Conditions**	测试仪器:DSC 坩埚:40 μl 铝坩埚,密封。 样品制备:原样品。 测试: 对于 A、B、C:由 30 ℃升温至 100 ℃,由 100 ℃降温至 30 ℃。第 2 次升温由 30 ℃至 175 ℃。试样量为5.057 mg。 对于 D、E:由 30 ℃升温至 175 ℃,由 175 ℃降温至 30 ℃。第 2 次升温由 30 ℃至 175 ℃。试样量为4.389 mg。各段升温速率均为 10 K/min,降温速率均为 5 K/min。 气氛:氮气,50 ml/min	**Measuring cell**:DSC **Pan**:Al 40 μl, hermetically sealed. **Sample preparation**:As received. **Measurement**: For A, B, C: Heating from 30 ℃ to 100 ℃. Cooling from 100 ℃ to 30 ℃. Second heating run from 30 ℃ to 175 ℃. The sample weight was 5.057 mg. For D, E: Heating from 30 ℃ to 175 ℃. Cooling from 175 ℃ to 30 ℃. Second heating run from 30 ℃ to 175 ℃. The sample weight was 4.389 mg. The heating rate for all the steps was 10 K/min and the cooling rate was 5 K/min. **Atmosphere**:Nitrogen, 50 ml/min

解释 如图 3.31 所示,与实验达到的最高温度有关,可观察到不同的峰。

Interpretation Different peaks can be observed in figure 3.31, depending on the maximum temperature reached. The first peak

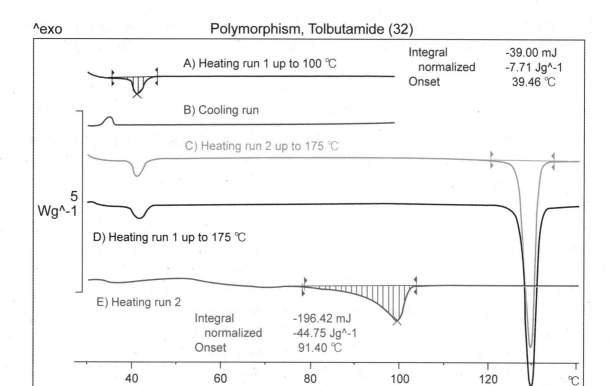

图 3.31 甲苯磺丁脲不同升降温的 DSC 曲线
Figure 3.31 DSC curves of tolbutamide at different heating runs and cooling runs

第 1 峰在约 40 ℃,不过,只有试样升温不高于 100 ℃,才是可逆的。这是由双向的固-固转变产生的。如果先是超过熔融温度(＞140 ℃),然后将试样降温,于是在第 2 次升温时可观察到一个具有宽熔融范围的新晶型。宽峰形表明有分解的杂质存在。

at about 40 ℃ is, however, only reversible if the sample is not heated above 100 ℃. It is caused by an enantiotropic solid-solid transition. If the melting temperature is first exceeded (＞140 ℃) and the sample afterwards cooled down, then a new modification can be observed with a broad melting range in the second heating run. The broad peak shape indicates the presence of impurities arising from decomposition.

计算
Evaluation

	第 1 个峰 First peak		中间峰 Middle peak		熔融峰 Melting peak	
	起始点 Onset ℃	ΔH J/g	起始点 Onset ℃	ΔH J/g	起始点 Onset ℃	ΔH J/g
A 第 1 次升温 A Heating run 1	39.5	7.7	—	—	—	—
B 降温 B Cooling run	36.2	6.2	—	—	—	—
C 第 2 次升温 C Heating run 2	39.5	7.5	—	—	126.4	0.9
D 第 1 次升温 D Heating run 1	39.3	7.1	—	—	126.9	91.5
E 第 2 次升温 E Heating run 2	—	—	91.4	44.8	—	—

结论 不同晶型样品的生成取决于热预处理,可用DSC清晰地区分不同晶型。

Conclusion Different modifications of the sample are formed depending on the thermal pretreatment. The modifications can be clearly distinguished with DSC.

3.29 退火处理丁基羟基茴香醚多晶型
Polymorphic Modifications by Annealing, Butylated Hydroxyanisole

样品 **Sample**	丁基羟基茴香醚 Butylated hydroxyanisole	
应用 **Application**	活性成分(口服抗糖尿病药) Active ingredient (oral antidiabetic)	
条件 **Conditions**	测试仪器:DSC	**Measuring cell**:DSC
	坩埚:40 μl 铝坩埚,密封。	**Pan**:Al 40 μl, hermetically sealed.
	样品制备:原样品或先经退火。	**Sample preparation**: As received or annealed beforehand.
	测试: A)以 2.5 K/min 由 30 ℃升温至 70 ℃ B)由 35 ℃升温至 60 ℃,在 60 ℃恒温 10min,然后降温至 30 ℃。 第 2 次升温以 2.5 K/min 由 30 ℃至 70 ℃。 气氛:氮气,50 ml/min	**Measurement**: A) Heating from 30 ℃ to 70 ℃ at 2.5 K/min B) Heating from 35 ℃ to 60 ℃, held isothermally for 10 min at 60 ℃, then cooling to 30 ℃ Second heating run from 30 ℃ to 70 ℃ at 2.5 K/min. **Atmosphere**:Nitrogen,50 ml/min

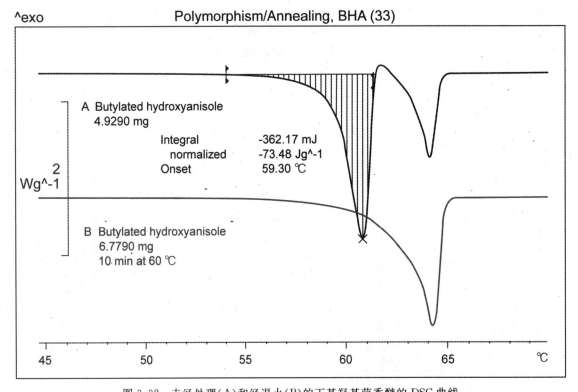

图 3.32 未经处理(A)和经退火(B)的丁基羟基茴香醚的 DSC 曲线

Figure 3.32 DSC curves of Butylated hydroxyanisoles unannealed (A) and annealed (B)

解释 如图 3.32 曲线 A 所示，未经处理的试样呈现带有一个结晶阶段的两个熔融峰。经退火处理的试样仅呈现一个熔融峰（曲线 B）。可在略高于第 1 个晶型的熔程起始点的温度退火，迫使试样结晶为另一种晶型。由"3.3 升温速率对丁基羟基茴香醚多晶型检测的影响"已知，第 1 种晶型的熔融热 ΔH 约为 100 J/g，而第 2 种晶型的熔融热（由本例）仅为 84.5 J/g。单向转变体系第 1 个峰的熔融热 ΔH 是小于第 2 个峰的。因而，本例中可能是双向互变体系。要搞清情况，可用热台显微镜在约 60 ℃ 进行测试。双向互变体系在两个吸热相之间是没有液相生成的。如果出现液相，则是单向转变体系，物质（曲线 B）不是 100% 结晶的。

计算 测定起始温度和焓变。括弧中表示因熔融与再结晶过程重叠而导致测定不确定的值。对试样 B 只表示第 2 次升温。

Interpretation The untreated sample shows two melting peaks with a recrystallization step, as shown by curve A in figure 3.32. The annealed sample exhibits just one melting peak (curve B). Annealing at a temperature slightly above the onset of the melting range of the first modification forces the sample to crystallize to the other form. As known from "3.3 Influence of Heating Rate on the Detection of Polymorphism, Butylated Hydroxyanisole", the heat of fusion ΔH of the first modification is about 100 J/g and that of the second form (from this example) only 84.5 J/g. With monotropic systems, the heat of fusion ΔH of the first peak is smaller than that of the second peak. Consequently in this case it could be an enantiotropic system. In order to clarify the situation, a measurement at about 60 ℃ with the hot stage microscope should be performed. In an enantiomeric system no liquid phase is formed between the two endotherms. If a liquid phase appears, then this is a monotropic system and the substance (curve B) was not 100% crystalline.

Evaluation The onset temperatures and the enthalpy changes are determined. The values that can not be determined with certainty because of the overlap of the melting and recrystallization processes are shown in brackets. In the case of sample B only the second heating run is shown.

试样 Sample	起始点 Onset ℃	ΔH J/g	起始点 Onset ℃	ΔH J/g
A（未处理） A (untreated)	59.3	(73.5)	63.0	(27.0)
B（在 60 ℃ 10 min） B (10 min at 60 ℃)	—	—	62.7	84.5

结论 物质的贮存条件或工艺条件不适当，可造成晶型改变，尤其是当转变温度接近贮存或加工温度时。

Conclusion Unfavorable storage or processing conditions of a substance can lead to crystal modification changes, especially when the transition temperature lies close to the storage or processing temperature.

3.30 硬脂酸镁的 DSC "指纹"
DSC 'Fingerprint', Magnesium Stearate

样品 硬脂酸镁

Sample	Magnesium stearate		
应用	非活性成分（片剂生产润滑剂）		
Application	Inactive ingredient (lubricant for production of tablets)		
条件	测试仪器：DSC	**Measuring cell**：DSC	
Conditions	坩埚：40 μl 铝坩埚，密封。	**Pan**：Al 40 μl, hermetically sealed.	
	样品制备：原样品。	**Sample preparation**：As received.	
	测试：	**Measurement**：	
	以 10 K/min 由 30 ℃升温至 200 ℃	Heating from 30 ℃ to 200 ℃ at 10 K/min	
	气氛：氮气，50 ml/min	**Atmosphere**：Nitrogen, 50 ml/min	

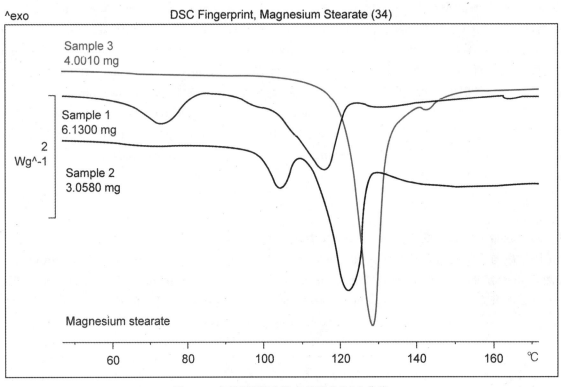

图 3.33　不同硬脂酸镁多晶型的 DSC 曲线

Figure 3.33　DSC curves of different polymorphic forms of magenesium stearate

解释　如图 3.33 所示，硬脂酸镁呈现若干个多晶型。此外，根据来源不同，还存在与游离脂肪酸的不同混合物。因此，指定单个晶型常常并不容易。样品还可生成含结晶水的晶型。

Interpretation　Magnesium stearate exhibits several polymorphic forms as shown in figure 3.33. In addition, depending on the origin, there are different mixtures with free fatty acids. Assignment to single modification is therefore often not so easy. The samples can also form modifications containing water of crystallization.

结论　DSC 能测定不同的多晶型，曲线用作不同硬脂酸镁质量的指纹。

Conclusion　DSC allows the determination of different polymorphic modifications. The curves are used as fingerprints for the different qualities of magnesium stearate.

3.31 左旋聚丙交酯的多晶型　Polymorphism, L-Polylactide

样品	左旋聚丙交酯		
Sample	L-Polylactide		
应用	非活性成分		
Application	Inactive ingredient		
条件	测试仪器：DSC	**Measuring cell**：DSC	
Conditions	坩埚：40 μl 铝坩埚，密封。	**Pan**：Al 40 μl, hermetically sealed.	
	样品制备：原样品。	**Sample preparation**：As received.	
	测试：	**Measurement**：	
	以 10 K/min 由 30 ℃升温至 300 ℃	Heating from 30 ℃ to 300 ℃ at 10 K/min	
	气氛：氮气，50 ml/min	**Atmosphere**：Nitrogen, 50 ml/min	

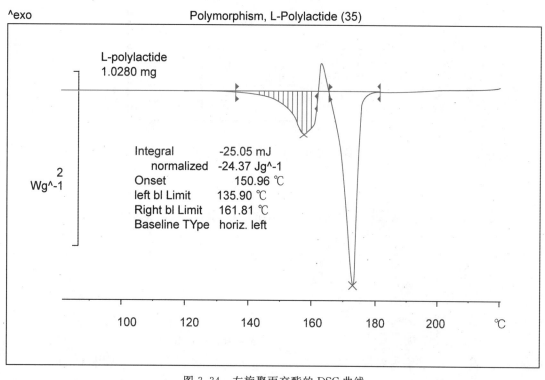

图 3.34　左旋聚丙交酯的 DSC 曲线
Figure 3.34　DSC curve of L-polylactide

解释　图 3.34 所示 DSC 曲线为在熔融态经历转变的物质所具有的典型形状。两个峰的分离情况取决于晶型的结晶速率。尽管分离良好，测得的第 1 种晶型的熔融热也是偏低的，这是因为：第一已存在未知数量的高熔融晶型，第二高熔融晶型的放热结晶是与第 1 种晶型的熔融同时发生，第三会发生固-固转变。

Interpretation　The form of the DSC curve shown in Figure 3.34 is typical for substances that undergo transformations in the melt. A separation of the two peaks depends on the crystallization rates of the modifications. In spite of the good peak separation, the measured heat of fusion of the first modification is too low：firstly, because an unknown quantity of the high-melting modification is already present, secondly, because the exothermic crystallization of the high-melting form takes place at the same time as the melting of the first form, and thirdly, because a solid-solid

第 2 峰通常偏小,因为在动态升温时转变即结晶是不完全的。所以,在计算表格中将两个熔融热都用括弧表示。

要产生高温晶型,可用"3.29 退火处理丁基羟基茴香醚多晶型"中描述的相似方法,将试样在 160 ℃退火。

计算 要正确计算熔融峰,必须应用几种不同类型的基线。对第 1 峰用"左水平"类型,而对第 2 峰用"右水平"类型。

transition can occur. The second peak is usually too small, because the transformation or crystallization is incomplete when heating dynamically. For this reason both heats of fusion are displayed in brackets in the evaluation table.

In order to produce the high temperature form, the sample could be annealed at 160 ℃ in a similar way to that described in "3.29 Polymorphic Modifications by Annealing, Butylated Hydroxyanisole".

Evaluation Several different baseline options have to be applied in order to evaluate the melting peak correctly. For the first peak the type 'horizontal left' is used and for the second peak the type 'horizontal right'.

	起始点 Onset ℃	ΔH J/g
第 1 峰 First peak	151.0	(24.4)
第 2 峰 Second peak	168.4	(58.9)

结论 本例说明聚合物也会呈现多晶型。

Conclusion The example demonstrates that polymers can also exhibit polymorphism.

3.32 磺胺吡啶的多晶型 Polymorphism, Sulfapyridine

样品	磺胺吡啶
Sample	Sulfapyridine
应用	活性成分
Application	Active ingredient
条件	测试仪器:DSC
Conditions	坩埚:40 μl 铝坩埚,密封。
	样品制备:原样品。
	测试:
	由熔体骤冷,然后以 5 K/min 由 40 ℃升温至 200 ℃
	气氛:氮气,50 ml/min

Measuring cell:DSC
Pan:Al 40 μl, hermetically sealed.
Sample preparation:As received.
Measurement:
Shock-cooled from the melt. Then heating from 40 ℃ to 200 ℃ at 5 K/min
Atmosphere:Nitrogen, 50 ml/min

解释 图 3.35 为磺胺吡啶的 DSC 曲线。

在进行动态测试前,将试样骤冷(快速冷却)至玻璃化转变温度 T_g 以下,常常是检测多晶型的最好方法。根据 Osward 说法,如果存在亚稳

Interpretation Figure 3.35 shows the DSC curve of Sulfapyridine.

Shock-cooling (rapidly cooling) the sample to a temperature below the glass transition temperature T_g before performing a heating run is often the best method of detecting polymorphism. According to Oswald, the metastable phase (B, if one exists at

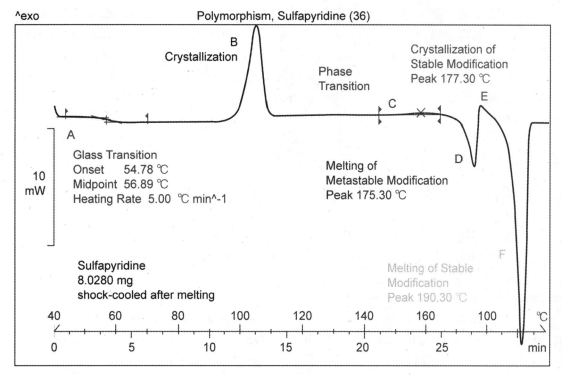

图 3.35 磺胺吡啶的 DSC 曲线
Figure 3.35 DSC curve of sulfapyridine

相 B,则会在玻璃化转变 T_g(A)以上结晶。继续升温时,先经单向转变(C,固-固转变),然后亚稳相熔融(D),由液相结晶为稳定晶型(E),最终稳定晶型熔融(F)。

由其他 DSC 测试得知,磺胺吡啶呈现更多的晶型。

all) crystallizes above the glass transition T_g(A). This is transformed on further heating first monotropically (C, solid-solid transition) and then, after the melting of the metastable phase (D), by crystallization from the liquid phase (E) to the stable modification. This then finally melts (F).

It is known from other DSC measurements, that sulfapyridine exhibits further modifications.

计算
Evaluation

	起始点或峰 Onset or Peak ℃	ΔH J/g	效应 Effect
A	54.8	—	玻璃化转变 glass transition
B	99.9	64.9	冷结晶 cold crystallization
C	150.1	(2.4)	单向转变 monotropic transition
D	175.3	(24.4)	尚存在亚稳相熔融 melting of the metastable phase still present
E	177.3	(2.4)	由熔体稳定相结晶 crystallization of the stable phase from the melt
F	190.3	(109.1)	稳定相的熔融 melting of the stable phase

结论 物质呈现多晶型的可能性,最好的研究方法是先非常仔细地使

Conclusion The possibility of a substance exhibiting polymorphism is best investigated by first of all melting it most carefully

之熔融(在氮气中快速熔融),然后立即将其骤冷至玻璃化温度以下(经验规则：T_g 约为 T_f 的 3/4,温度单位均为开尔文。)。之后,将试样在 DSC 样品皿内缓慢升温,便可观察到不同的晶型。

(rapidly, under nitrogen) and then immediately shock-cooling it under its glass transition temperature (rule of thumb: T_g is roughly 3/4 T_f, both temperatures are in Kelvin). Afterwards, the different modifications can be observed by slowly heating the sample in the DSC cell.

3.33 一水葡萄糖的假多晶型
Pseudopolymorphism, Glucose Monohydrate

样品	一水和无水葡萄糖	
Sample	Glucose, monohydrate and the anhydrous form	
应用	非活性成分	
Application	Inactive ingredien	
条件	测试仪器：DSC 和 TGA	Measuring cell: DSC and TGA
Conditions	坩埚:40 μl 或 100 μl 铝坩埚,盖钻孔。	Pan: Al 40 μl or 100 μl, with pierced lid.
	样品制备:原样品。	Sample preparation: As received.
	DSC 测试:	DSC Measurement:
	以 20 K/min 由 30 ℃升温至 250 ℃。	Heating from 30 ℃ to 250 ℃ at 20 K/min.
	TGA 测试:	TGA Measurement:
	以 20 K/min 由 30 ℃升温至 300 ℃。	Heating from 30 ℃ to 300 ℃ at 20 K/min
	气氛:	Atmosphere:
	氮气,DSC:50 ml/min,TGA:80 ml/min	Nitrogen, DSC: 50 ml/min, TGA: 80 ml/min

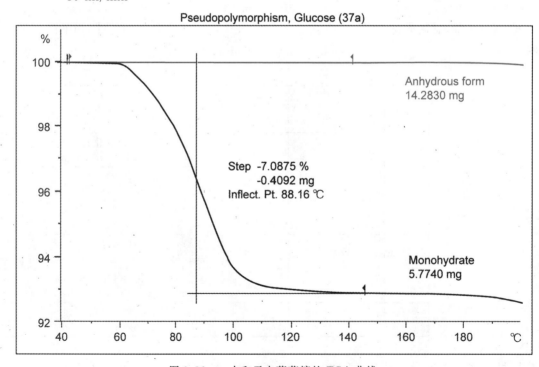

图 3.36 一水和无水葡萄糖的 TGA 曲线
Figure 3.36 TGA curves of Glucoses of monohydrate and the anhydrous form

解释 葡萄糖可以一水形式或无水形式存在。图 3.36 中的 TGA 曲线表明，两种形式完全不同。无水形式不呈现失水，而一水形式可观察到失水，约 7% 台阶的测量结果略低于化学计量预测值 9.1%。

Interpretation Glucose can exist as the monohydrate or in the anhydrous form. The TGA curves in figure 3.36 show that the two forms differ quite distinctly from one another. The anhydrous form shows no loss of moisture, whereas with the monohydrate the elimination of water can be observed. This results in a step of about 7% and is slightly below the value of 9.1% expected stoichiometrically.

计算 测定失重。

Evaluation The weight loss is determined.

试样 Sample	台阶 Step %	计算范围 Eval. Range ℃
一水葡萄糖 Monohydrate	7.1	42～145
无水葡萄糖 Anhydrous form	<0.1	42～145

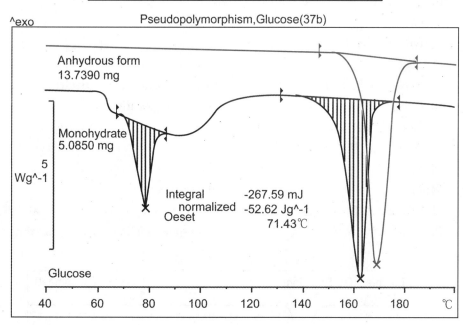

图 3.37 一水和无水葡萄糖的 DSC 曲线

Figure 3.37 DSC curves of Glucoses of monohydrate and the anhydrous form

解释 如图 3.37 中 DSC 曲线所示，无水形式只呈现 161 ℃ 左右的熔融过程。由失水转化成无水形式，一水葡萄糖呈现很宽的峰。然后约在 158 ℃ 熔融。熔融峰越小，熔点就比无水形式的越低，表明结晶不完全。

Interpretation As the DSC curves in figure 3.37 show, the anhydrous form exhibits only the melting process at about 161 ℃. The monohydrate shows a very broad peak resulting from the elimination of water and the transformation to the anhydrous form. This then melts at about 158 ℃. The smaller heat of fusion and lower melting point in comparison with the anhydrous form, indicate an incomplete crystallization.

计算 测定起始温度和峰面积。

Evaluation The onset temperatures and peak areas are determined.

试样 Sample	第 1 峰 First peak		第 2 峰 Second peak	
	起始点 Onset ℃	ΔH J/g	起始点 Onset ℃	ΔH J/g
一水葡萄糖 Monohydrate	71.4	52.6	153.9	162.4
无水葡萄糖 Anhydrous form	—	—	160.9	198.7

结论 DSC 和 TGA 曲线能鉴别葡萄糖的一水形式和无水形式。

Conclusion The DSC and the TGA curves allow the identification of the monohydrate and anhydrous forms.

3.34 布洛芬(异丁苯丙酸)的光学纯度 Optical Purity, Ibuprofen

样品 布洛芬(异丁苯丙酸),外消旋体和(+)对映体
Sample Ibuprofen, racemate and (+) enantiomer
应用 活性成分
Application Active ingredient
条件 测试仪器:DSC
Conditions 坩埚:40 μl 铝坩埚,密封。
样品制备:
原样品。将混合物溶解在少量甲醇中,在 6 ℃放置若干天使其慢慢结晶。
测试:
以 5 K/min 由 30 ℃升温至 110 ℃
气氛:氮气,50 ml/min

Measuring cell:DSC
Pan:Al 40 μl, hermetically sealed.
Sample preparation:
As received. The mixtures are dissolved in a little methanol and left to crystallize slowly for several days at 6 ℃.
Measurement:
Heating from 30 ℃ to 110 ℃ at 5 K/min
Atmosphere:Nitrogen, 50 ml/min

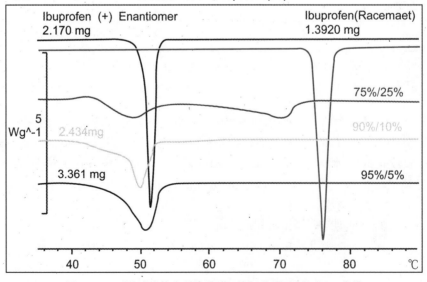

图 3.38 不同对映体含量布洛芬(异丁苯丙酸)的 DSC 曲线

Figure 3.38 DSC curves of Ibuprofens with different enantiomer concentration

解释 图 3.38 为不同对映体含量的布洛芬（异丁苯丙酸）的 DSC 曲线。

同分异构体为化学组成相同但结构不同的物质。对映体为呈现光学活性的同分异构体。如果（＋）与（－）对映体以相同比例存在于一个晶体中，则称为外消旋体。外消旋物和纯对映体有不同的晶体结构、不同的溶解性和熔点，因此可用 DSC 来区别。本例中，（＋）对映体为药物活性形式，它比外消旋体的溶解性更大、熔点更低。这在药物制剂配方中会有影响，例如，如果外消旋体与对映体相互能够取代。

计算 测定熔融峰的起始温度和熔融热。绘制相图要求用熔融起始温度。

图 3.39 所示相图（液相线）表示熔融起始温度与（－）对映体含量的关系。含量 0% 即 100%（＋）对映体，50% 等于外消旋体，100% 为纯（－）对映体。相图的右边（含量 50% 至 100%）由左边部分镜像生成。

Interpretation Figure 3.38 shows the DSC curves of Ibuprofens with different enantiomer concentration.

Isomers are substances with the same chemical composition but different structure. Enantiomers are isomers which show optical activity. If the (＋) and (－) enantiomers are present in the same ratio in a crystal, then it is known as a racemate. Racemates and pure enantiomers have different crystal structures, different solubilities and melting points and can therefore be distinguished from each other by DSC. The (＋) enantiomer is, in this case, the pharmacologically active form. Compared with the racemate, it has a greater solubility and a lower melting point. This can have consequences in the formulation of pharmaceutical preparations, if for example, the racemate and enantiomer should be able to be replaced by one another.

Evaluation The onset temperatures and heats of fusion of the peaks are determined. The onset temperatures are required for constructing the phase diagram.

The phase diagram (liquidus curve) as shown in figure 3.39 shows the onset temperatures as a function of the (－) enantiomer concentration. A concentration of 0% means 100% (＋) enantiomer, 50% corresponds to the racemate, and 100% is pure (－) enantiomer. The right section of the diagram (concentration 50% to 100%) was obtained by forming the mirror image of the left part.

试样 Sample	起始点 Onset ℃	ΔH J/g	起始点 Onset ℃	ΔH J/g
外消旋体 Racemate	—	—	74.4	121.8
（＋）对映体 （＋）Enantiomer	50.4	86.6	—	—
95%对映体/5%外消旋体 95% enantiomer/5% racemate	47.5	81.7	—	—
90%对映体/10%外消旋体 90% enantiomer/10% racemate	47.6	77.4	—	—
75%对映体/25%外消旋体 7% enantiomer/25% racemate	43.6	35.7	63.7	39.4

注：25%外消旋体相当于 12.5%（－）对映体。
Note：25% racemate corresponds to 12.5% (－) enantiomer.

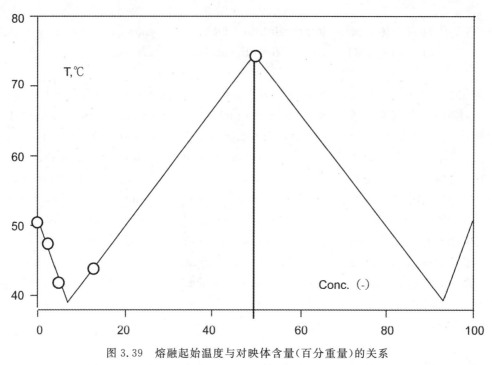

图 3.39 熔融起始温度与对映体含量(百分重量)的关系
Figure 3.39 The onset temperature as a function of the enantiomer concentration (weight%)

结论 DSC 能将外消旋物与纯的对映体区分开,可检测(+)对映体中混杂有外消旋物,通过熔融峰变宽和熔点降低来观察。

然而,DSC 无法将纯的(+)对映体与(−)对映体区分开来,因为它们的熔点相同。不过,可将(+)对映体加入未知对映体中:熔融峰不变表示未知物中也存在于(+)对映体中。

Conclusion DSC can distinguish between racemates and the pure enantiomers. A contamination of the (+) enantiomer with racemate can be detected. This is observed as a broadening of the melting peak and a depression of the melting point.

DSC can not, however, distinguish between the pure (+) and (−) enantiomers because they have the same melting points. One can, however, add the (+) form to the unknown enantiomer: an unchanged melting peak shows that the unknown substance is also present in the (+) form.

3.35 对羟基苯甲酸及其酯的纯度测定 (DSC 法和 HPLC 法)
Purity using DSC and HPLC, 4-Hydroxybenzoic Acid and its Esters

样品　　　　1 对羟基苯甲酸 4.28 mg
　　　　　　　2 对羟基苯甲酸丁酯 3.61 mg
　　　　　　　3 对羟基苯甲酸丙酯 3.75 mg
　　　　　　　4 对羟基苯甲酸乙酯 4.54 mg
　　　　　　　5 对羟基苯甲酸甲酯 4.22 mg
　　　　　　　(R = H、丁基、丙基、乙基、甲基)

Sample　　　1 4-Hydroxybenzoic acid 4.28 mg
　　　　　　　2 Butyl-4-hydroxybenzoate 3.61 mg
　　　　　　　3 Propyl-4-hydroxybenzoate 3.75 mg
　　　　　　　4 Ethyl-4-hydroxybenzoate 4.54 mg
　　　　　　　5 Methyl-4-hydroxybenzoate 4.22 mg

	(R=H, Butyl, Propyl, Ethyl, Methyl)	
应用 **Application**	非活性成分 Inactive ingredient	
条件 **Conditions**	测试仪器：DSC	**Measuring cell**：DSC
	坩埚：40 μl 铝坩埚，密封。	**Pan**：Al 40 μl, hermetically sealed.
	样品制备：在坩埚中向下压实试样。	**Sample preparation**：Sample pressed down in the pan.
	测试：1 由 205 ℃升温至 220 ℃ 2 由 50 ℃升温至 80 ℃ 3 由 85 ℃升温至 100 ℃ 4 由 105 ℃升温至 120 ℃ 5 由 120 ℃升温至 130 ℃ 所有实验的升温速率均为 1 K/min。	**Measurement**：1 Heating from 205 ℃ to 220 ℃ 2 Heating from 50 ℃ to 80 ℃ 3 Heating from 85 ℃ to 100 ℃ 4 Heating from 105 ℃ to 120 ℃ 5 Heating from 120 ℃ to 130 ℃ Heating rate for all experiments was 1 K/min.
	气氛：空气，静止环境，无流动	**Atmosphere**：Air, stationary environment, no flow

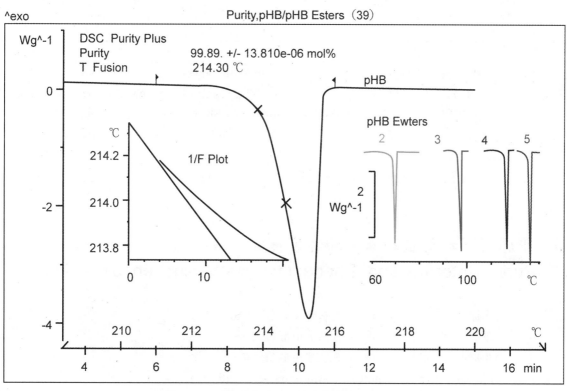

图 3.40 对羟基苯甲酸及其酯的熔融 DSC 曲线

Figure 3.40 DSC melting curves of 4-Hydroxybenzoic Acid and its Esters

解释 试样对羟基苯甲酸及其酯的熔融曲线如图 3.40 所示。按 van't Hoff 纯度测定 DSC 法计算了试样的纯度，图中以对羟基苯甲酸（PHB）作为例子，显示其详细计算。

Interpretation The melting curves of the samples of 4-Hydroxybenzoic Acid and its Esters are shown in figure 3.40. The purity of the samples are evaluated using the DSC method for purity determination according to van't Hoff. As an example, the detailed evaluation for 4-hydroxybenzoic acid (PHB) is described in the figure.

计算 用"直线"基线在峰高的10%至50%间计算纯度。此外,用HPLC法测定了各物质的纯度。

Evaluation Evaluation of the purity between 10% and 50% of the peak height using the baseline type 'line'. In addition, the purity of the substances was determined with HPLC.

试样 Sample	对羟基苯甲酸 4-Hydroxy-benzoic-acid	丁酯 Butyl-ester	丙酯 Propyl-Ester	乙酯 Ethyl-ester	甲酯 Methy-ester
熔点 Melting point ℃	214.30	69.10	96.43	115.79	125.75
熔融热 Heat of fusion J/g	244.13	159.87	151.83	160.21	173.46
DSC法纯度 Purity DSC Mol%	99.90	99.76	99.65	99.88	99.89
HPLC法纯度 Purity HPLC %	100.0	99.96	99.94	99.87	99.98

结论 所测试的物质呈现很高的纯度,可认为由 DSC 和 HPLC 获得的结果在误差范围内是相同的。
在对比两种方法得到的结果时,须考虑到 HPLC 法的结果是以质量分数表示的,而 DSC 法的结果为摩尔分数。

Conclusion The substances investigated show a very high degree of purity. The results from DSC and from HPLC can be regarded as identical within the limits of accuracy.
When comparing the results of the two methods, one must allow for the fact that the HPLC results are expressed in units of weight percent, and the DSC results in mole percent.

3.36 非那西汀+ 对氨基苯甲酸纯度测定
Purity Determination, Phenacetin + 4-Aminobenzoic Acid

样品	非那西汀 + 0.7 mol% 对氨基苯甲酸(PABS) 非那西汀 + 2.0 mol% 对氨基苯甲酸(PABS) 非那西汀 + 5.0 mol% 对氨基苯甲酸(PABS)
Sample	Phenacetin + 0.7 mole% 4-Aminobenzoic acid (PABS) Phenacetin + 2.0 mole% 4-Aminobenzoic acid (PABS) Phenacetin + 5.0 mole% 4-Aminobenzoic acid (PABS)
应用	作为纯度测定标准
Application	As a standard for purity determination
条件 **Conditions**	测试仪器:DSC　　**Measuring cell**:DSC 坩埚:40 μl 铝坩埚,密封。　　**Pan**:Al 40 μl, hermetically sealed. 样品制备:原试样。　　**Sample preparation**:As received. 测试:　　**Measurement**: 以 5 K/min 由 100 ℃升温至 160 ℃　　Heating from 100 ℃ to 160 ℃ at 10 or 5 K/min 以 1.25 K/min 由 110 ℃升温至 150 ℃　　Heating from 110 ℃ to 150 ℃ at 2.5 or 1.25 K/min

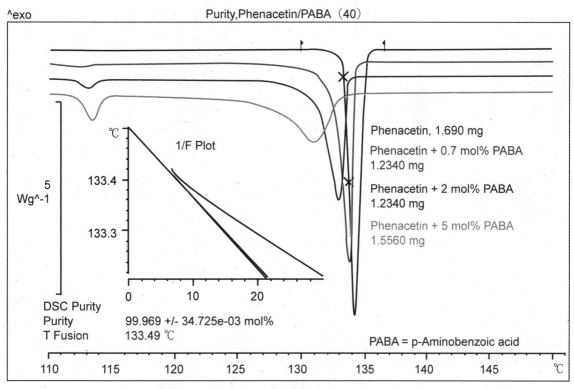

图 3.41　纯非那西汀及其受污染试样的熔融 DSC 曲线
Figure 3.41　DSC melting curves of pure and contaminated samples of phenacetin

解释　图 3.41 为纯非那西汀及其受污染试样的 DSC 熔融曲线。随着杂质增加,熔融峰变宽且移至较低温度。此外,约 114 ℃ 的共熔峰越来越明显。纯度测定基于 van't Hoff 方程,它表明熔点降低与熔体杂质分数成正比。图 3.41 中表示熔融曲线平衡熔融温度对熔融分数(F)的倒数 1/F 作图。在考虑了热阻校正和熔融峰面积修正后,由它生成的线性关系可测定纯度。

Interpretation　Figure 3.41 shows the melting curves of the pure sample and the contaminated samples of phenacetin. The melting peak becomes broader and shifts to lower temperature with increasing impurity. In addition, the eutectic peak at about 114 ℃ becomes increasingly noticeable. The purity determination is based on the van't Hoff equation, which states that the melting point depression is proportional to the mole fraction impurity. The diagram in figure 3.41 showing the equilibrium melting temperature from the melting curve plotted against the reciprocal of the fraction melted (F) is known as the 1/F plot. After allowing for the heat of fusion of the eutectic (linearization correction), a linear relationship is generated from which the purity can be determined.

计算　下面给出的每个试样的纯度是依照 van't Hoff 方法在所示升温速率测定的。在峰高的 10% 至 50% 范围内计算熔融曲线。

Evaluation　The purity of each of the samples given below was determined at the heating rates indicated, according to the method of van't Hoff. The melting curve was evaluated in the range from 10% to 50% of the peak height.

试样 Sample	DSC 纯度,Purity DSC,Mol%				
	标准值 Certified value	1.25 K/min	2.5 K/min	5 K/min	10 K/min
纯非那西汀 Phenacetin pure	99.9±0.1	99.97	99.98	99.96	99.97
+0.7mol% PABS	99.3±0.2	99.39	99.29	99.34	99.50
+2.0mol% PABS	98.0±0.2	98.35	98.39	98.49	98.40
+5.0mol% PABS	94.9±0.2	98.30	97.07	96.79	97.04

结论 DSC 纯度测定可快速得到结果。然而，只有当满足一定条件时，才能应用该方法。尤其是，只有相对纯的物质（>98%）才应用该技术来表征。

在开发一个纯度测定方法时，必须考虑许多不同的参数，包括颗粒大小、试样量、升温速率以及选择用于计算的数据点。颗粒大小的变化可引起熔融峰变宽或不规则，因为经过较大颗粒的热流需要较长时间。大的试样量也会有相同的影响，因为试样厚度增大。另一方面，陷入细粉末中的空气可造成各个晶粒间热传递的延迟。

Conclusion The DSC purity determination gives rapid results. The method may only be applied, however, if certain conditions are fulfilled. In particular, only relatively pure substances (>98%) should be characterized with this technique.

A number of different parameters must be considered when developing a method for purity determination. These include particle size, sample weight, heating rate and choice of the data points to be used for the calculation. A variation in the particle size can cause broader or irregular melting peaks, since the heat flow through larger crystals requires more time. A large sample weight has the same effect because of the increased sample thickness. On the other hand, trapped air in fine powders can lead to a delayed transfer of heat between the individual crystals.

3.37 胆甾醇的纯度和重结晶
Purity and Recrystallization, Cholesterol

样品 Sample	>99%和>95%不同纯度的胆甾醇 Cholesterol, various degrees of purity >99% and >95%	
条件 Conditions	测试仪器：DSC 坩埚：40 μl 铝坩埚,密封。 样品制备： 用特氟隆棒将试样压入坩埚。通过由甲醇重结晶3次将标称为>95%纯度的试样纯化。 测试： 以 2 K/min 由 120 ℃升温至 170 ℃ 气氛：空气,静止环境,无流动	**Measuring cell**：DSC **Pan**：Al 40 μl, hermetically sealed. **Sample preparation**： The sample was pressed down in the pan with a teflon rod. The sample with the stated purity of >95% was purified by recrystallizing 3 times from methanol. **Measurement**： Heating from 120 ℃ to 170 ℃ at 2 K/min **Atmosphere**：Air, stationary environment, no flow

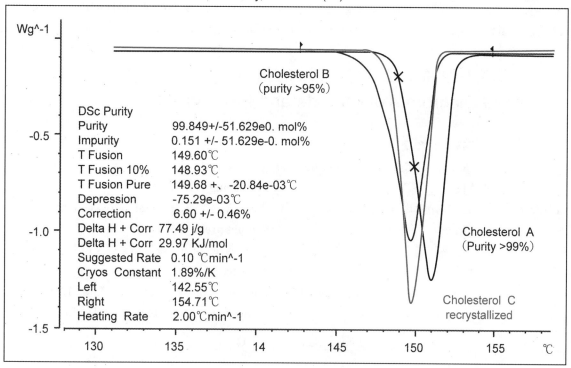

图 3.42 三种不同胆甾醇试样的熔融 DSC 曲线
Figure 3.42 DSC melting curves of three Cholesterol samples

解释 图 3.42 为三种不同胆甾醇试样的熔融 DSC 曲线。按 van't Hoff 的纯度测定方程计算纯度。作为例子，选用试样之一显示其详细计算。

所测定的纯度几乎相同，好于制造商的指标（＞95％的试样也是这样）。就此方法而言，应该注意，DSC 只能测量共熔杂质，而其他杂质是检测不到的。

Interpretation Figure 3.42 shows the DSC melting curves of the three Cholesterol samples. The purities are evaluated using the purity determination according to van't Hoff. As an example, the detailed evaluation is displayed for one of the samples.

The purities determined are almost identical and are (also for the sample ＞95％) better than the manufacturer's specification. In this connection, it should be noted that DSC can only measure eutectic impurities, so that other impurities remain undetected.

计算 用"直线"基线在峰高的 10％ 至 50％ 间计算纯度。

Evaluation Evaluation of the purity between 10％ and 50％ of the peak height with baseline type 'line'.

试样 Sample	纯度 Purity Mol％	起始点 Onset ℃	重量 Weight mg
胆甾醇 A（纯度＞99％） Cholesterol A（purity ＞99％）	99.85	148.8	11.08
胆甾醇 B（纯度＞95％） Cholesterol B（purity ＞95％）	99.51	147.5	10.80
胆甾醇 C（重结晶试样） Cholesterol C（recrystallized）	99.87	148.4	5.41

结论 通过重结晶可纯化产品。如果只是涉及共熔杂质，则可用 DSC 纯度测定法定量测定纯度。

Conclusion Products can be purified by recrystallization. The degree of purification can be quantified with a DSC purity determination, provided that we are dealing with a eutectic impurity.

3.38 甲苯磺丁脲和聚乙二醇 6000 的相图
Phase Diagram, Tolbutamide and PEG 6000

样品 **Sample**	甲苯磺丁脲和聚乙二醇 6000 Tolbutamide(TBA) and PEG 6000 as well as mixtures	
条件 **Conditions**	测试仪器：DSC	**Measuring cell**：DSC
	坩埚：40 μl 铝坩埚，密封。	**Pan**：Al 40 μl, hermetically sealed.
	样品制备：	**Sample preparation**：
	原样品或两组分的物理混合物。	As received or physical mixtures of both components.
	测试：	**Measurement**：
	以 10 K/min 由 30 ℃升温至 175 ℃	Heating from 30 ℃ to 175 ℃ at 10 K/min
	气氛：氮气，50 ml/min	**Atmosphere**：Nitrogen, 50 ml/min

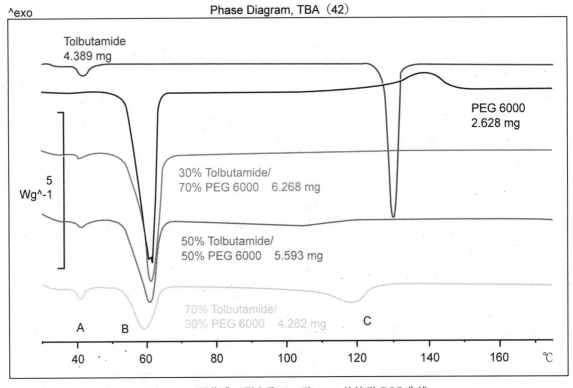

图 3.43　甲苯磺丁脲和聚乙二醇 6000 的熔融 DSC 曲线
Figure 3.43　DSC melting curves of samples of Tolbutamide(TBA) and PEG 6000

解释 甲苯磺丁脲与聚乙二醇 6000 生成固体分散体，但不是固溶液。在固体溶液中，各组分形成确定熔程的混合晶体。本例中，如图 3.43 所示，甲苯磺丁脲与聚乙二醇

Interpretation Tolbutamide forms a solid dispersion with PEG 6000, but not a solid solution. In a solid solution the components form mixed crystals with a definite melting range. In this case, as shown in figure 3.43, tolbutamide and PEG 6000 form a eutectic with a composition of 30% TBA and 70% PEG 6000

6000生成组成为30％甲苯磺丁脲和70％聚乙二醇的共熔体,熔点与所用聚乙二醇的相同。

约39 ℃的峰A为甲苯磺丁脲的固-固转变。在60 ℃,可观察到峰B。如果甲苯磺丁脲大于30％,则过量的甲苯磺丁脲发生熔融(峰C)。

计算 由熔融峰计算起始温度和熔融热 ΔH。

如图3.44所示的相图,可由起始温度对甲苯磺丁脲(TBA)含量(重量百分数)来绘制。TBA的固-固转变发生在约39 ℃,这里可称为晶型A和B。高于54 ℃,取决于TBA的含量,或者为液相,或者为固相B。虚线是外推的。要确认这些外推线绘制完整相图,需要用不同组成的混合物作进一步测量。

with the same melting point as the PEG 6000 used.

The peak A at about 39 ℃ corresponds to a solid-solid transition of tolbutamide. At 60 ℃ peak B can be observed. Excess TBA melts (peak C) if more than 30％ TBA is present.

Evaluation The onset temperatures and the heats of fusion ΔH are evaluated from the melting curves.

The following phase diagram as shown in figure 3.44 can be constructed by plotting the onset temperature against the concentration of TBA (in weight percent). The solid-solid transition of TBA occurs at about 39 ℃, named here as modifications A and B. Above 54 ℃, either the liquid or the solid phase B is present depending on the TBA concentration. The dotted lines are extrapolations. Additional measurements with mixtures of different composition would have to be performed in order to confirm these extrapolations and to construct a complete phase diagram.

试样 Sample	峰 A Peak A		峰 B Peak B		峰 C Peak C	
	起始点 Onset ℃	ΔH J/g	起始点 Onset ℃	ΔH J/g	起始点 Onset ℃	ΔH J/g
TBA 100％	39	8	—	—	127	93
TBA 90％/PEG 10％	39.1	7.4	53.7	19.8	121.1	75
TBA 70％/PEG 30％	39.3	5.8	53.2	51.6	109	48
TBA 50％/PEG 50％	39.4	3.9	54.3	53.9	81.6	22.8
TBA 30％/PEG 70％	39.5	1.5	55.9	138.8	104.8	—
TBA 10％/PEG 90％	39.5	0.3	55.3	166.7	116.4	—
PEG 100％	—	—	55	178	128	—

结论 用DSC可测定双组分混合物的相图。

Conclusion Phase diagrams of binary mixtures can be determined by DSC.

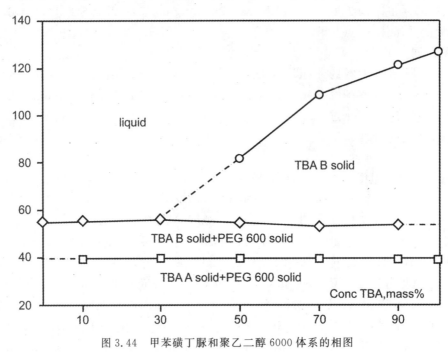

图 3.44　甲苯磺丁脲和聚乙二醇 6000 体系的相图

Figure 3.44　Phase diagram for the system of Tolbutamide(TBA) and PEG 6000

3.39 对羟基苯甲酸甲酯和对羟基苯甲酸的共熔体组成
Eutectic Composition, Methyl-4-Hydroxybenzoate and 4-Hydroxybenzoic Acid

样品 **Sample**	对羟基苯甲酸甲酯(MpHB)、对羟基苯甲酸(pHB)及其不同混合物 Methyl-4-hydroxybenzoate (MpHB), 4-Hydroxybenzoic acid (pHB) as well various mixtures	
条件 **Conditions**	测试仪器：DSC 坩埚：40 μl 铝坩埚，密封。 样品制备： 在研钵中研磨制备试样。试样在坩埚内压紧。 测试： 以 5 K/min 由 115 ℃升温至 135 ℃ 气氛：氮气，50 ml/min	**Measuring cell**：DSC **Pan**：Al 40 μl, hermetically sealed. **Sample preparation**： Mixtures prepared by grinding in a mortar. The samples were pressed down in the pan. **Measurement**： Heating from 115 ℃ to 135 ℃ at 5 K/min **Atmosphere**：Nitrogen, 50 ml/min

解释　图 3.45(a)和(b)所示为纯对羟基苯甲酸甲酯(MpHB)的尖锐熔融峰以及不同 MpHB/pHB 混合物的共熔峰。pHB 含量少于 10% 的混合物呈现较小的共熔峰以及过量 MpHB 温度降低的熔融峰。纯 pHB 的熔点约为 124 ℃。

Interpretation　Figure 3.45(a) and Figure 3.45(b) show the sharp melting peak of the pure MpHB, as well as the eutectic peaks of the different mixtures of MpHB/pHB. The mixtures with a pHB content of less than 10% show smaller eutectic peaks as well as the depressed melting peak of the excess MpHB. The melting point of pure pHB is about 214 ℃.

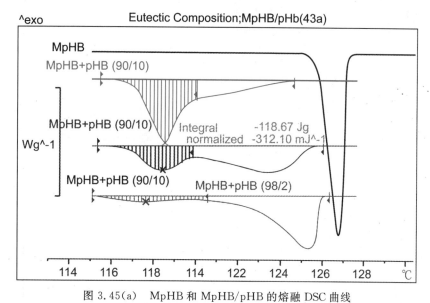

图 3.45(a) MpHB 和 MpHB/pHB 的熔融 DSC 曲线

Figure 3.45(a) DSC melting curves of MpHB and MpHB/pHB

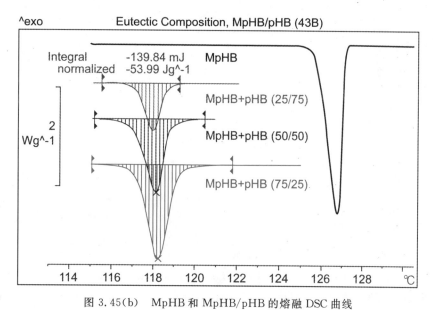

图 3.45(b) MpHB 和 MpHB/pHB 的熔融 DSC 曲线

Figure 3.45(b) DSC melting curves of MpHB and MpHB/pHB

计算 对所有混合物,由共熔峰计算熔融热。高 pHB 含量(>10%)的混合物用"直线"基线计算。其他不对称峰用"左水平"基线计算其部分面积。为了测定共溶体,绘制了混合物熔融热与组成关系图,见图 3.46。两条直线的交点,即共熔点,表征共熔体。共熔体组成为 14.16 摩尔% pHB 和 85.84 摩尔% MpHB,共熔温度为 117.4 ℃。

Evaluation The heats of fusion are calculated from the eutectic peaks for all the mixtures. The mixtures with high pHB content (>10%) are evaluated using the baseline type 'line'. The other assymetric peaks are evaluated using the baseline type 'horizontal left' and calculating the partial integral. In order to determine the eutectic, a diagram is constructed showing the heats of fusion versus the composition of the mixtures, as shown in Figure 3.46. The point of intersection of the two straight lines, the eutectic point, characterizes the eutectic. The eutectic composition is 14.16 mole% pHB and 85.84 mole% MpHB and the eutectic temperature 117.4 ℃.

MpHB/pHB mol%	熔融热 ΔH Heat of fusion ΔH J/g	试样量 Sample weight mg
98/2	22.5	8.05
95/5	58.2	3.42
90/10	118.9	2.63
75/25	144.3	2.96
50/50	97.1	2.88
25/75	54.0	2.59

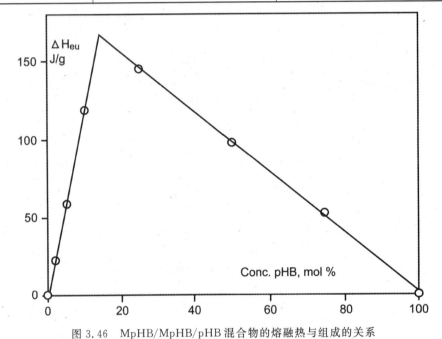

图 3.46　MpHB/MpHB/pHB 混合物的熔融热与组成的关系

Figure 3.46　Heats of fusion versus the composition of MpHB/MpHB/pHB mixures

结论　共熔峰熔融热的计算可精确测定本双组分体系的共熔体组成。该方法的准确性取决于测量点数和精确计算峰的可能性。建议通过测量相同组成混合物的 DSC 曲线来检查计算得到的共熔体组成。

Conclusion　The evaluation of the heat of fusion of the eutectic peak allows an exact determination of the eutectic composition of this two-component system. The accuracy of the method depends on the number of points measured and the possibility of a precise evaluation of the peaks. It is recommended, that the calculated eutectic composition be checked by measuring a DSC curve of a mixture of the same composition.

3.40　药物活性物质的 TGA-MS 溶剂检测
Solvent Detection by means of TGA-MS, Pharmaceutically Active Substance

样品　在有机溶剂中重结晶得到的药物活性物质
Sample　Pharmaceutically active substance, recrystallized in an organic solvent
应用　活性成分。

Application	Active ingredient.		
条件 Conditions	测试仪器：TGA + 质谱仪	**Measuring cell**：TGA + MS	
	坩埚：70 μl 氧化铝坩埚	**Pan**：Alumina 70 μl	
	样品制备：原样品	**Sample preparation**：As received	
	测试：	**Measurement**：	
	以 10 K/min 由 30 ℃升温至 350 ℃， 作空白曲线修正。	Heating from 30 ℃ to 350 ℃ at 10 K/min, blank curve corrected.	
	气氛：氮气，20 ml/min	**Atmosphere**：Nitrogen, 20 ml/min	

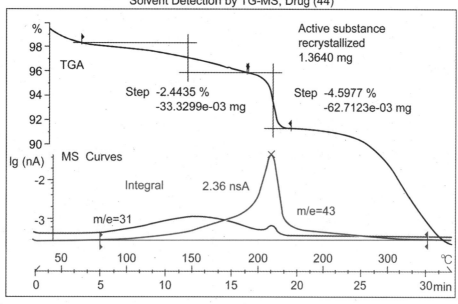

图 3.47 药物活性物质的 TGA 和 MS 曲线

Figure 3.47 TGA and MS curves of the Pharmaceutically active substance

解释 在药物的合成和之后的提纯/重结晶过程中经常要使用各种溶剂。药物中残留溶剂的存在会影响其性能。因此，必须鉴定有关的溶剂，使其含量尽可能低。

图 3.47 中的 TGA 曲线呈现若干个失重台阶。在 250 ℃的最后台阶，试样开始分解。70～240 ℃范围的两个台阶表明，在加热过程中失去水或溶剂。同步记录的 MS 离子曲线证实，失重台阶为甲醇(m/z 31)和丙酮(m/z 43,丙酮的主要碎片离子)。甲醇在很宽的温度范围内释放，相比而言，丙酮消除的温度范围要窄得多。这表明丙酮键合更坚固，可能为溶剂化物。

Interpretation Different solvents are often used both in the synthesis of a pharmaceutical substance and afterward for its purification/recrystallization. The presence of residual amounts of solvents in a substance can influence its properties. The solvents concerned must therefore be identified and their concentration kept as low as possible.

The TGA curve shown in figure 3.47 exhibits several mass loss steps. In the final step above 250 ℃, the substance begins to decompose. The two steps in the range 70－240 ℃ indicate that moisture or solvents are lost through heating. The simultaneously recorded MS ion curves confirm that the mass loss steps correspond to methanol (m/z 31) and acetone (m/z 43, the main fragment ion of acetone).

The methanol is released over a wide temperature range. In comparison, the acetone is eliminated in a much narrower temperature range. This indicates that the acetone is more firmly bound, possibly as a solvate.

计算 Evaluation		台阶高度 Step height %	计算范围 Evaluation range ℃	溶剂 Solvent
	台阶 1 Step 1	2.5	70—190	甲醇 methanol
	台阶 2 Step 2	4.5	190—225	丙酮,甲醇 acetone, methanol
	台阶 3 Step 3	开始分解 beginning of decomposition	—	—

结论 在适用情况下,TGA 和 MS 的联用可表征溶剂化物,有可能将吸收的残留溶剂与以某种方式键合更牢的溶剂分子分辨开来。

Conclusion In favorable cases the combination of TGA and MS allows the characterization of solvates and makes possible the distinction between absorbed residual solvent and solvent molecules that are more firmly bound in some way.

3.41 不同水分含量的油包水乳膏的定量分析
Quantification, O/W Creams with Different Water Content

样品 含如下组分(%)的乳膏:
Sample Creams with the following compositions (in%):

样品 Sample	A	B	C	D	E
油醇 Oleyl alcohol	10	10	—	—	—
乳化油 Mygliol 812	15	15	—	—	—
十六烷醇 Cetyl alcohol	4	4	44	22	5.8
十八烷醇 Stearyl alcohol	4	4	44	22	5.8
单硬脂酸甘油酯 Glyceryl monostearate	2	—	—	—	2.9
鲸蜡硬脂基磺酸钠 Lanette E	1	1	11	5.5	1.5
苯甲醇 Benzyl alcohol	1	—	—	—	—
丙二醇 Propylene glycol	5	5	—	—	—
NaOH	0.02	0.02	—	—	—
柠檬酸 Citric acid	0.05	0.05	—	—	—
水 Water	57.93	59.93	<1	50	83.8

应用 制造乳膏的原料。
Application Basic materials for the manufacture of creams.

条件	测试仪器：TGA	Measuring cell：TGA
Conditions	坩埚：	Pan：
	100 μl 铝坩埚，盖钻直径 1mm 的孔	Al 100 μl, the lid was pierced, diameter of hole 1 mm
	样品制备：原样品	Sample preparation：As received
	测试：	Measurement：
	以 2 K/min 由 20 ℃升温至 200 ℃，作空白曲线修正。	Heating from 20 ℃ to 200 ℃ at 2 K/min, blank curve corrected.
	气氛：氮气，80 ml/min	Atmosphere：Nitrogen，80 ml/min

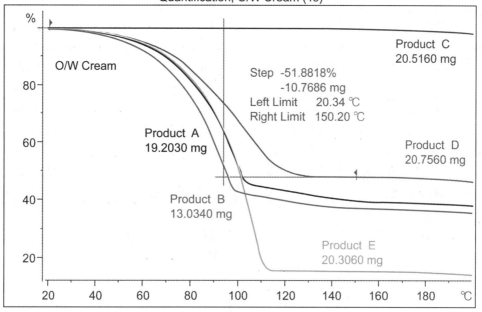

图 3.48　不同油包水乳膏试样的 TGA 曲线

Figure 3.48　TGA curves of samples of the different O/W cream

解释　如图 3.48 所示，除了试样 C，所有乳膏在 30 ℃至 140 ℃间呈现显著的失重。这是由挥发性成分的蒸发导致的，首先是水。

Interpretation　As shown in figure 3.48, all the creams, with the exception of sample C, show a significant weight loss in the range 30 ℃ to 140 ℃. This is caused by the evaporation of volatile components, above all water.

计算　用 TGA 曲线计算各个台阶。

Evaluation　The individual steps are evaluated using the TGA curves.

试样 Sample	台阶高度 Step height %	拐点 Inflection ℃	计算范围 Evaluation range ℃
A	60.6	96.1	20−161
B	62.7	92.7	20−153
C	0.53	—	20−146
	51.8	91.4	20−149
	84.4	97.8	20−131

结论　测量结果稍高于给定的水含

Conclusion　The results lie slightly higher than the values expected

量值,可解释为有其他挥发性成分蒸发,导致失重大于预计值。

from the specified water content. This can be explained by assuming that other volatile components evaporate and in this way cause a weight loss that is greater than expected.

3.42 一水茶碱的定量分析
Quantification, Theophylline Monohydrate

样品	一水茶碱和无水茶碱	
Sample	Theophylline monohydrate and the anhydrous form	
条件	测试仪器:DSC 和 TGA	**Measuring cell**: DSC and TGA
Conditions	DSC 坩埚:40 μl 铝坩埚,密封。	**Pan DSC**: Al 40 μl, hermetically sealed.
	TGA 坩埚:100 μl 铝坩埚,盖钻孔。	**Pan TGA**: Al 100 μl, with pierced lid.
	样品制备:	**Sample preparation**:
	原样品,或在相对湿度53%和92%贮存的一水茶碱。	As received, or storage of the monohydrate at a rel. humidity of 53% and 92%.
	DSC 测试:	**DSC Measurement**:
	以 20 K/min 由 20 ℃升温至 300 ℃。	Heating from 20 ℃ to 300 ℃ at 20 K/min.
	TGA 测试:	**TGA Measurement**:
	以 20 K/min 由 30 ℃升温至 300 ℃,空白曲线修正。	Heating from 30 ℃ to 300 ℃ at 20 K/min, blank curve corrected.
	气氛:	**Atmosphere**:
	氮气,DSC:50 ml/min,TGA:80 ml/min	Nitrogen, DSC: 50 ml/min, TGA: 80 ml/min

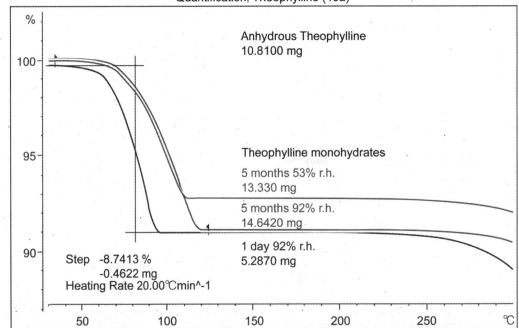

图 3.49 在不同相对湿度贮存的一水茶碱和无水茶碱的 TGA 曲线
Figure 3.49 TGA curves of theophylline monohydrate stored under different relative humidity conditions and the anhydrous form

解释 图 3.49 中 TGA 曲线表明, **Interpretation** It can be shown from the TGA curves in figure

结晶水的百分含量取决于贮存条件。在相对湿度大于92%贮存时，一水茶碱保持稳定，失重约为9%，与化学计量计算预计的相同。在较低湿度下贮存时，发生部分结晶水失去。为了对比，示出无水茶碱的TGA曲线，如同预计的，不呈现失水。

TGA 计算 由 TGA 台阶测定失水。

3.49 that the percentage content of water of crystallization is dependent on the storage conditions. The monohydrate remains stable when stored at a relative humidity greater than 92% and shows a loss of about 9% water, as expected from a stoichiometric calculation. Storage at lower humidity results in the partial loss of water of crystallization. For comparison, the curve of the anhydrous form is shown. As expected, this shows no weight loss.

Evaluation TGA The loss of moisture is determined from the TGA steps.

样品 Sample	TGA 台阶 TGA Step %	计算范围 Evaluation range ℃	拐点 Inflection pt ℃
无水茶碱 Anhydrous form	<0.1	30-150	—
一水茶碱 Monohydrate	8.7	30-150	—
相对湿度 92%1 天 1 Day 92% r. h.	8.7	32-130	82.6
相对湿度 92%5 个月 5 Months 92% r. h.	8.9	34-124	99.6
相对湿度 53%5 个月 5 Months 53% r. h.	7.2	33-123	95.7

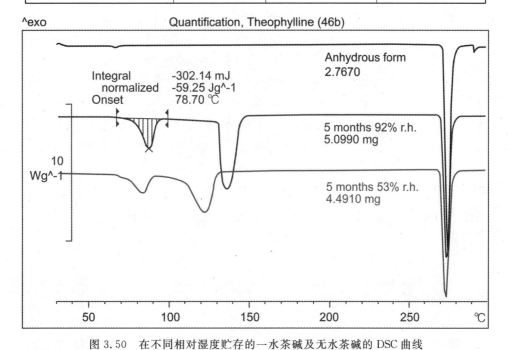

图 3.50 在不同相对湿度贮存的一水茶碱及无水茶碱的 DSC 曲线

Figure 3.50 DSC curves of Theophylline monohydrate stored under different relative humidity conditions and the anhydrous form

解释 从图 3.50 的 DSC 曲线可观

Interpretation In the DSC curves shown in Figure 3.50, it can

察到，在较高相对湿度下贮存，第 1 峰（失去结晶水）的热焓显著增大。基线漂移的第 2 峰是由水蒸气压力导致的坩埚胀破产生的。因此，在数据处理时，这些峰的 ΔH 值加上了括弧。无水茶碱是通过水的蒸发得到的，约于 272 ℃ 熔融。含无水茶碱的坩埚也会胀破，但在较高温度（约 290 ℃）。

DSC 计算 计算了所有峰的起始温度和积分。

be seen that storage at higher relative humidity results in a significant increase in the enthalpy of the first peak (loss of water of crystallization). The second peak followed by a baseline shift is caused by the pan bursting as a result of the pressure of the water vapor. In the evaluation, the ΔH values for these peaks are therefore put in brackets. The anhydrous form is produced through the evaporation of the water. This melts at about 272 ℃. The pan containing the anhydrous form of theophylline also bursts but at higher temperature (about 290 ℃).

Evaluation DSC The onset temperatures and integrals of all the peaks are evaluated.

试样 Sample	峰 1 Peak 1		峰 2 Peak 2		峰 3 Peak 3	
	起始点 Onset ℃	ΔH J/g	ΔH J/g	起始点 Onset ℃	ΔH J/g	
无水茶碱 Anhydrous form	63.1	2.1	—	272.7	169.4	
相对湿度 53% 5 个月 5 Months 53% r. h.	73.1	40.3	(121.6)	270.9	138.5	
相对湿度 92% 5 个月 5 Months 92% r. h.	78.7	59.3	(168.8)	271.3	141.0	

结论 含结晶水形式的稳定性可用 DSC/TGA 测定。关于这种水合形式稳定性的信息，对于药物生产很重要，因为会显著影响药物制剂的含量。尤其对于即使小剂量也呈高活性的高活性药，影响会很大。

Conclusion The stability of a form containing water of crystallization can be determined by DSC/TGA. Information concerning the stability of such hydrated forms is of great importance for the manufacture of pharmaceutical products, because the content of a pharmaceutical preparation can be affected to significant degree. This can have serious consequences especially with highly active drugs that show high activity even in small doses.

3.43 Alcacyl 中活性物质的测定
Determination of an Active Substance, Alcacyl

样品 Sample	Alcacyl 中的卡巴匹林钙（乙酰水杨酸钙盐＝Ca-ASA） Calcium carbasalate (calcium salt of acetylsalicylic acid＝Ca-ASA) in Alcacyl
应用 Application	止痛药 Analgesic
条件 Conditions	测试仪器：DSC 和 TGA　　　　　Measuring cell：DSC and TGA DSC 坩埚：40 μl 铝坩埚，盖钻孔。　Pan DSC：Al 40 μl, with pierced lid. TGA 坩埚：100 μl 铝坩埚，盖钻孔。Pan TGA：Al 100 μl, with pierced lid.

样品制备：原样品。	Sample preparation: As received.
DSC 测试：	DSC Measurement:
以 10 K/min 由 30 ℃升温至 300 ℃。	Heating from 30 ℃ to 300 ℃ at 10 K/min.
TGA 测试：	TGA Measurement:
以 10 K/min 由 30 ℃升温至 300 ℃。	Heating from 30 ℃ to 300 ℃ at 10 K/min.
气氛：	Atmosphere:
氮气，DSC：50 ml/min，TGA：80 ml/min	Nitrogen, DSC: 50 ml/min, TGA: 80 ml/min

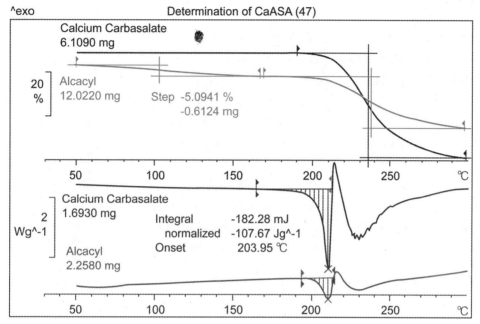

图 3.51　Alcacyl 药片和纯卡巴匹林钙的 DSC 和 TGA 曲线

Figure 3.51　DSC and TGA curves of Alcacyl tablet and pure Calcium carbasalate

解释　可用 DSC 来检测活性成分，测定熔点和熔融热，如图 3.51 所示。不过，需有纯物质作参比。如果活性成分与非活性成分间没有相互作用，则通过药片的重量和制造商提供的信息就有可能进行分析。Alcacyl 的 TGA 曲线呈现两个台阶，而 Ca-ASA 只有一个。将 Alcacyl 的第 2 台阶（=－24％）与 Ca-ASA 的分解台阶（=－49％）对比表明，其失重与活性物质分解所预计的值几乎完全相等。

TGA 计算　由于试样在熔融时分解，所以用"左水平"基线进行 DSC 峰的积分。

Interpretation　The active ingredient can be detected with DSC and the melting point and the heat of fusion can be measured, as shown in figure 3.51. This assumes, however, that the pure substance is available for reference purposes. Analysis is possible using the tablet's weight and information from the manufacturer, provided that no interaction occurs between the active and inactive ingredients. The TGA curve of Alcacyl shows two steps whereas that of Ca-ASA just one. A comparison of the second step of Alcacyl (=－24％) with the decomposition step of Ca-ASA (=－49％), shows that the weight loss corresponds almost exactly to the loss expected by decomposition of the active substance.

Evaluation TGA　Since the samples decompose on melting, the integration of DSC peak is performed using the baseline type 'horizontal left'.

样品 Sample	DSC 起始点 Onset DSC ℃	DSC ΔH J/g	TGA 台阶1 TGA Step 1 %	TGA 台阶2 TGA Step 2 %
卡巴匹林钙 Ca carbasalate	203.9	107.3	—	49.0
Alcacy 药片 Alcacy tablet	203.0	28.2	5.1	24.3

8片药片的平均重量为1.09008 g，卡巴匹林钙含量为528 mg，等于活性成分含量值为48.4%。对于Alcacyl的DSC分析，由48.4%活性成分含量，预期热量变化为51.6 J/g（107.3 J/g×0.484＝51.6 J/g）。对于TGA分析，台阶应为约23.7%（49%卡巴匹林钙失重 x 活性含量 0.484＝23.7）。

TGA实际测量结果与预期值吻合，而DSC测量并不吻合，其中原因是难于将熔融峰与分解峰清晰分开。

结论 如果片剂中的活性成分与非活性成分不发生相互作用，则活性成分的分析在原则上是可能的。也可以用熔融热估算含量。本例中，TGA是更准确的方法，因为同时发生的分解过程干扰了DSC分析。

The mean value of the weight of 8 tablets was 1.09008 g and the Ca-Carbasalate content was 528 mg. This corresponds to a value of 48.4% for the content of the active ingredient. For the DSC analysis of Alcacyl a heat change of 51.6 J/g (107.3 J/g×0.484 = 51.6 J/g) is expected for the active ingredient content of 48.4%. For the TGA analysis the step should correspond to about 23.7% (49% weight loss of the Ca-Carbasalate x active ingredient content 0.484 = 23.7%).

While the actual TGA measuring results agree with the values expected, this is not the case with the DSC measurement. The reason here is the difficulty of clearly separating the melting peak from the decomposition peak.

Conclusion An analysis of the active ingredient is in principle possible, provided that no interaction occurs between the active and the inactive ingredients of the tablet. The content can also be estimated using the heat of fusion. In this case the TGA is the more accurate method, since the decomposition process that occurs at the same time interferes with the DSC analysis.

4 热分析在食品工业的应用
Applications of Thermal Analysis in the Food Industry

4.1 热分析食品应用一览表　Application Overview Food

表格所示为可用热分析测试的食品的效应和性能。黑点表示较重要的应用。

The table shows the effects and properties of food products that can be investigated by thermal analysis. The more important applications are marked with a black spot.

	DSC	TGA	TMA	DMA
熔融和结晶 Melting and crystallization	●		○	○
干燥、蒸发、升华、解吸附 Drying, evaporation, sublimation, desorption	●	●		
多晶型 Polymorphism	●		○	○
玻璃化转变 Glass transition	●		●	●
比热容 Specific heat capacity	●			
热稳定性 Thermal stability	○	●		
氧化稳定性 Oxidation stability	●	○		
焓变、反应焓 Enthalpy changes, reaction enthalpies	●			
化学反应、变性 Chemical reaction, denaturation	●	●		
组分分析、纯度 Compositional analysis, purity	●	●		
塑料包装材料的表征 Characterization of plastic packging materials	●	●	●	●

4.2 食品工业与热分析　Food Industry and Thermal Analysis

4.2.1 食品工艺中的反应和相　Reactions and Phases in Food Technology

热（即温度）是用于食品加工中的主要控制因素之一。DSC 能表征这些过程（例如沸腾或冷冻），并能借助于温度程序，测量其他的物理或化学性能（例如淀粉的溶胀或包装薄膜的组成）。

食品典型的热过程：
● 低热灭菌/消毒（奶制品/罐头食品/饮料）
● 烘烤、沸腾（面包、糕饼和甜点、肉、熟食）

Heat (i. e. temperature) is one of the main controlling factors used in food processing. DSC allows the characterization of these processes (e. g. boiling or freezing) and, with the aid of temperature programs, the determination of additional physical or chemical properties (e. g. the swelling of starch or the composition of packaging films).

Typical Thermic Food Processes：
● Pasteurization/sterilization (milk products/canned goods, beverages)
● Baking, boiling (bread, cakes and pastries, meat, ready cooked meals)

- 干燥（蛋白质、素菜）
- 冷却（脂肪、熟食）
- 冷冻（肉、素菜）

热过程引起的变化：
- 相（蒸发、熔融；结晶、冷冻）
- 构形（脂肪晶体改变、变性）
- 化学反应，诸如氧化

- Drying (proteins, vegetables)
- Cooling (fats, ready cooked meals)
- Freezing (meat, vegetables)

Changes Caused by Thermal Processes
- Phase state (vaporization, melting; crystallization, freezing)
- Conformation (fat crystal modification, denaturation)
- Chemical reaction such as oxidation

4.2.2 食品中主要成分 DSC 检测一览表
List of DSC Investigations of the Main Components in Foods

成分 Component	效应 Effect	课题 Topics	行业 Industry
蛋白质 Proteins	变性 Denaturation	● 检测热历史（热作用熟化） Detection od the 'thermic history' (thermic influence ageing) ● 组分、分级 Composition, fractionation ● 与脂质、碳水化合物的作用 Interaction with lipids, carbohydrates	奶、肉和素菜蛋白加工业 Milk, meat and vegetable protein processig industry
淀粉 Starch	溶胀 Swelling	● 溶胀度 Degree of swelling ● 溶胀温度 Swelling temperatures ● 凝胶化 Gelatinization	面包、糕饼甜点和马铃薯加工业 Bread, cakes and pastries and potato processing industry
	回生 Retrogradation	● 面包陈放 Staling of bread	
脂质 Lipids	熔融 结晶 晶体转变 Melting Crystallization Crystal transitions	● 固体脂肪含量（SFI） Solid fat fraction (SFI) ● 熔融和结晶动力学 Melting and crystallization kinetics ● 多晶型 Polymorphism ● 贮存行为 Storage behavior ● 组分 Composition	糖果盒食用油加工业 Confectionery and edible oil processing industry
水 Water	冷冻 Freezing	● 不结冻水含量 Nonfreezing fraction of water	速冻产品加工业 Deep-frozen products processing industry
塑料包装材料 Plastic packaging	熔融 Melting	● 复合薄膜的鉴别 Identification of composite films	所有食品行业 Entire food industry

4.2.3 蛋白质 Proteins

根据其在营养学和生理学方面的意义及其在功能性开发方面的作用,蛋白质与脂肪和碳水化合物一起,是最重要的食品成分之一。其对食品的加工(例如乳化)和消费品表征(感官性能)很重要。

蛋白质是由多达 20 种不同氨基酸组成的长链分子,它们不仅分子组成不同,而且空间结构也不一样,这是技术性功能开发的决定性因素。

关于蛋白质的变性

食品加工中最常发生的蛋白质反应是所谓的蛋白质变性。这是一个描述众多化学反应(分子内键例如氢键或硫醚键的断裂、肽链展开并聚集形成杂乱的凝块等)的通用术语。这些反应可由热引发,也可由清洁剂、酸、碱处理或由机械应力引起。一旦引发,反应将平行或连续快速进行,结果实际上不可能单独跟踪,因而只可区分为"原生"状态和"变性"状态。不过,变性并不均匀发生在整个物质,因而可将变性物质部分与非变性部分区分开来。

总体上,变性是吸热反应,在食物中产生复杂的、通常为不可逆的系统变化。反应焓一般在 100 kJ/mol~400 kJ/mol 蛋白质。

DSC 是根据通常发生在 40 ℃ 至 100 ℃ 间的吸热信号来检测变性,如果在该温度范围出现其他可逆的化学反应或相转变,则可通过同一试样第 2 次测量时信号消失(不可逆性)来鉴别蛋白质变性。这主要用于食品加工的实践性应用。对蛋白质变性基本性质的早期热力学研究,采用 DSC 曾发现其稀溶液中的可逆反应。

由测定的反应焓(等于 DSC 峰积分)可获得关于试样中变性蛋白质

Proteins, together with fats and carbohydrates, are among the most important food constituents on account of their significance in nutrition and physiology and their role in the development of functional properties. These are important for the processing of the food (e.g. emulsification) and the characterization of the consumable product (sensory properties).

Proteins are long chain molecules made up of as many as twenty different amino acids. They differ not only in their molecular composition but also in their spatial structure, a factor that is decisive in the development of the technofunctional properties.

On the Denaturation of Proteins

The protein reaction that occurs most frequently in food processing is the so-called protein denaturation. This is a collective term that describes a multitude of chemical reactions (intramolecular bonds such as hydrogen or disulfide bonds are broken, the peptide chain unfolds and aggregates to form a disorganized coagulum etc.). These reactions can be induced thermicly and also by treatment with detergents, acids and bases or by mechanical stress. Once initiated, the reactions run in parallel or consecutively so rapidly that it is virtually impossible to follow them individually and hence only the states 'native' and 'denatured' can be distinguished. However, the denaturation does not occur homogeneously throughout the entire material, so that a distinction can be made between the fractions of denatured materials.

Overall, denaturation is an endothermic reaction and occurs in foods as a complex, usually irreversible system change. The reaction enthalpy is normally between 100－400 kJ/mol protein.

DSC detects the denaturation as an endothermic signal, which is usually in the range 40 ℃ to 100 ℃. Should other reversible chemical reactions or phase transitions appear in this temperature range, the protein denaturation can be identified by the absence of a signal (irreversibility) in a second measurement of the same sample. This applies primarily in the practice-oriented applications of food processing. Earlier DSC investigations of a fundamental nature on the thermodynamics of protein denaturation also showed reversible reactions in dilute solutions.

The measured reaction enthalpy corresponding to the integral of the DSC peak provides information on the fraction of denatured

含量的信息。参考值为原生蛋白质的变性焓。另一方面,经加工的蛋白质的反应焓通常较低,因为部分蛋白质已在加工期间变性。

加工参数对蛋白质质量的影响

变性通常在营养可利用方面并不起决定性作用(例如,尽管煮熟的鸡蛋的蛋白质已经变性,但人仍可利用),蛋白质的技术性功能(诸如与水键合的能力、乳化功能、发泡稳定或香味络合)的开发,非常依赖于其"完整无损"。变性了的蛋白质实际上是不可能再溶解的。变性程度可直接与许多技术性功能关联,因而可在加工过程中测量食品加工参数对蛋白质的影响。结合 DSC 方法的另一个优势,即测定不同相中和不同浓度(所有固/液和液/固分散液以及乳化液)的样品,使 DSC 成为用于包括中间产物的多阶段过程中蛋白质加工方法过程优化的一个合适工具。

在蛋白质产品加工过程中,蛋白质部分通常发生改变,部分是有意的,但是仍应部分保留原生结构。测试中会显示两个常规加工参数的影响,即影响蛋白质质量的温度/时间效应和 pH 值。

文献中描述的在蛋白质方面的更多应用

- 热处理(冷冻、加热、干燥)、pH 值、水含量、添加盐、糖、矿物质或清洁剂对鱼肉、牛奶和乳清、鸡蛋、牛肉、谷物、油籽、豆类的影响

- 溶解性、肌理、乳化能力或表面活性等质量指标的 DSC 测定
- 加工条件(油菜和大豆蛋白的离析、腊肠加工)对产品质量的影响
- 猪肉屠宰后阶段的评判、新鲜肉的质量评估
- 在烘烤过程中蛋白质与淀粉和蛋的蛋白质等其他食物成分的相互作用

protein in the sample. The reference value is the denaturation enthalpy of native protein. Processed protein on the other hand usually has lower reaction enthalpies since fractions of the protein may have already been denatured during the treatment.

Influence of Process Parameters on Protein Quality

Whereas denaturation frequently does not play a decisive part in the nutritional utilizability (e. g. although the proteins of a cooked egg are denatured, they can be utilized by man), the development of the technofunctional properties of the proteins such as the ability to bond with water, emulsifying power, foam stabilization or aroma complexation depends greatly on their 'intactness'. Denatured protein is virtually impossible to redissolve. The degree of denaturation can be directly correlated with a number of technofunctional properties, the influence of process parameters on proteins in food can thus be determined in the processing. The combination of this fact with another advantage of the DSC method, namely the ability to measure samples in different phases and of different consistency (all solid/liquid and liquid/solid dispersions as well as emulsions), makes DSC a suitable tool for the process optimization of protein processing methods when used with intermediate products in a multi-stage process.

In the processing of protein products, the protein fraction is usually modified, in part intentionally, but the native structure should still be partially retained. The influence of two customary process parameters, namely the temperature/time effect and the pH value on the protein quality is shown in the measurements.

Additional Protein Applications Described in the Literature

- Influence of thermic treatment (freezing, heating, drying), the pH value, the water content, the addition of salts, sugars, minerals or detergents on the quality of the protein fractions of fish muscle, milk and whey, hen's eggs, bovine muscle, grain, oil seeds, beans
- Determination of quality criteria such as solubility, texture, emulsifying power, or surface activity by DSC
- Influence of the process conditions (isolation method for rape and soybean proteins, sausage processing) on the product quality
- Estimation of the post-mortem phase of pork, quality assessment of fresh meat
- Interaction of the proteins with other food components such as starch and egg proteins in baking processes

- 蛋白质变性反应动力学的测定（反应速率常数和活化能）
- Determination of the reaction kinetics of protein denaturation (reaction rate constant and activation energy).

4.2.4 碳水化合物　Carbohydrates

在碳水化合物领域，主要用 DSC 表征淀粉凝胶化和回生。面包烘烤、谷物产品挤制或酱汁稠化等过程的基础是淀粉的凝胶化。

用于研究淀粉性能的方法有许多（黏度测定、X 射线衍射、显微镜光学检测等），但 DSC 以应用范围广（其温度范围和水/淀粉比例范围均宽广）而著称，并且可测定转变焓。碳水化合物的热降解用热重分析仪研究。

In the field of carbohydrates, DSC is used primarily for characterization of the gelatinization and the retrogradation of starches. Processes such as the baking of bread, the extrusion of grain products or the thickening of sauces are based on the gelatinization of starches. Many methods are in use to investigate the properties of starches (viscosity measurement, x-ray diffraction, microscopic investigations etc.), but DSC is distinguished by a broad application range with regard to both the temperature range and the possible water/starch ratio. In addition, the enthalpy of transition can be measured. The thermal degradation of carbohydrates is investigated by thermogravimetry.

淀粉的凝胶化

加热时淀粉在水中的凝胶化是一系列形态变化的结果。在反应过程中，淀粉颗粒的结晶度下降，吸收热量，随着淀粉颗粒的溶胀淀粉的水合反应开始。进一步的吸水产生体积增大，直至颗粒取向结构的完全破坏。当水超过 60% 时就发生该过程。在热力学上，溶胀可看作是半结晶合成聚合物的熔融转变。

原生淀粉凝胶时，预期在 50 ℃至 80 ℃间出现吸热峰，依赖于淀粉的来源，转变焓为 7 J/g～11 J/g。

The Gelatinization of Starch

The gelatinization of starch in water on heating is the result of a number of conformational changes. In the course of the reaction the crystallinity of the starch granules is reduced, heat is absorbed and hydration of the starch sets in with swelling of the starch granules. Further water uptake leads to an increase in volume until the complete breakup of the oriented structure of the granules. This process occurs when water is available in excess of 60%. Thermodynamically the swelling can be viewed as the melting transition of a semicrystalline synthetic polymer.

In the gelatinization of native starches, an endothermic peak between 50 ℃ and 80 ℃ is expected. The enthalpy of transition is 7～11 J/g, depending on the origin of the starch.

在碳水化合物方面的更多应用：

- 脂质含量对淀粉溶胀的影响
- 例如在烘烤过程中淀粉与蛋白质的相互作用
- 改性淀粉（乙酰化、水解、预凝胶化）的表征
- 力学效应、热处理或加入添加剂（NaCl、蔗糖）对淀粉溶胀的影响
- 例如测量在贮存时间的回生过程（溶解淀粉分子的重结晶）的测定
- 反应过程模型和反应动力学的开发
- 用 TGA 测量热分解
- 对糖果和面条的热机械测量。

Additional Applications for Carbohydrates：

- Influence of the lipid content on starch swelling
- Interaction between starch and proteins, e.g. in the baking process
- Characterization of modified starches (acetylation, hydrolysis, pregelatinization)
- Influence of mechanical effects, thermic treatment or the addition of additives (NaCl, saccharose) on starch swelling
- Measurement of the retrogradation (recrystallization of dissolved starch molecules), i.e. determination of the storage time
- Development of models for the course of the reaction and the reaction kinetics
- Thermal decomposition by TGA
- Thermomechanical measurements on sweets and pasta.

4.2.5 脂肪和油 Fats and Oils

用DSC研究脂肪和油是在食品工业中最为广泛使用的常规应用之一。由于固体脂肪的多晶型结构，解释曲线需要凭经验了解存在多晶型和可能发生的转变过程。由于存在各种性能不同的晶体结构及其有时快速向其他晶型的转变，如欲获得再现性结果，则对实验条件的可靠性提出了很高的要求。

DSC的升温和降温速率影响脂肪的行为。结晶与熔融曲线是不一样的：降温时结晶条件对形成的晶型有决定性影响；升温熔融时，与升温速率有关，可能发生结构重排。升温时，还可获得与结晶历史和贮存条件有关的信息。脂肪的凝固性能对这类食物具有特别的技术意义。例如，可可脂的结晶对于巧克力质量是具有决定性重要作用的工艺步骤。就是在该步骤，巧克力产品获得符合需要的性能，例如光滑的表层、断裂强度和熔融行为（如"柔软性"）。

由于天然脂肪例如可可脂含有大量不同的甘油三酯，所以这些混合体系没有熔点，但是在一个温度范围熔融。须特别注意脂肪的多晶型，即相同组成的甘油三酯可能在晶格中排列不同。可可脂的主要成分是 POS（≈22%）、SOS（≈57%）和 POP（≈4%），式中 O 代表油酸、P 代表棕榈酸、S 代表硬脂酸。每一种甘油三酯，例如 POS，都呈现多晶型。含若干多晶型的甘油三酯混合物的熔融行为甚至更为复杂。在可可脂中，利用 X 射线衍射法至少可鉴别 γ、α、β'、β 4 种多晶型，热力学稳定性、熔点、熔融热和熔融膨胀性（由于液相密度较低导致熔融时的体积增大）依次增强。

The investigation of fats and oils with DSC is one of the most widely used routine applications in food processing. Because of the polymorphic structure of solid fats, the interpretation of the curves requires experience and knowledge of the modifications and the transition processes that may occur. The existence of various crystalline structures with different properties and their sometimes rapid conversion to other modifications places high demands on the reliability of the experimental conditions that are required to obtain reproducible results.

The heating and cooling rates of DSC influence the behavior of fats. Crystallization and melting curves are not identical: during cooling the crystallization conditions have a decisive influence on the modification formed; during heating and melting, rearrangements can occur, which are dependent on the heating rate. On heating, information is also obtained relating to the crystallization history and the storage conditions. The solidification properties of fats are of particular technological interest for the processing of such foodstuffs. For instance, the crystallization of cocoa butter is a process step of decisive importance for the quality of chocolate. It is only in this step that chocolate products obtain their desired properties such as glaze, breaking strength and melting behavior (e.g. 'softness').

Since natural fats such as cocoa butter contain numerous different triglycerides, these mixed systems do not have a melting point, but melt over a temperature range. Particular attention must be paid to polymorphism of the fats, i.e. triglycerides of the same composition can differ in their arrangement in the crystal lattice. The main constituents of cocoa butter are POS ($\approx 22\%$), SOS ($\approx 57\%$) and POP ($\approx 4\%$) where O means oleic, P is palmitic and S is stearic. Each triglyceride, e.g. POS, exhibits polymorphism. The melting behavior of mixtures of triglycerides containing several polymorphic forms is even more complex. In cocoa butter at least 4 polymorphs can be identified by x-ray diffraction. The modifications are designated γ, α, β', β. The thermodynamic stability, the melting point, the heat of fusion and the melting dilatation (a volume increase in melting due to the lower density of the liquid phase) increase in this order.

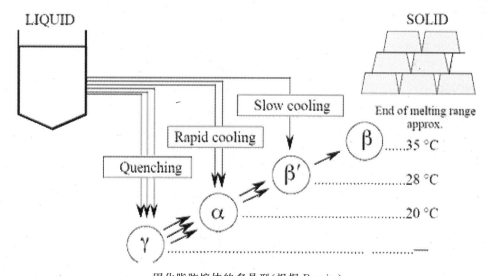

固化脂肪熔体的多晶型（根据 Baenitz）
Polymorphism of solidified fat melts (according to Baenitz)

无定形 γ-型，也称为玻璃态，由熔融脂肪骤冷得到。退火使该状态变化为 α-构型，然后最终变为稳定的 β-构型。通常，工业结晶的目标是获得尽可能高含量的稳定的高熔点 β-晶型。

β'-晶型是相对不稳定的，有时候出现直接的 α-β 重排。重排过程是发热的，而熔融当然是吸热的。以低升温速率可容易地检测到涉及不稳定中间相的重排。由于在通常的测试方法中无法自由选择结晶温度，所以显而易见的解决方案是用等温 DSC 操作作为常规方法（冷却至结晶温度，在结晶过程中保持温度不变）。

食用油和脂肪的自动氧化会对含脂肪产品的贮存和加工有负面影响。在室温时，氧化较慢，但在有氧高温下反应速率快速提高。150 ℃ 以上，例如在烘烤或煎炸时，则开始自动氧化；在更高温度，甚至发生脂肪的完全分解。

可用 DSC 测量脂肪的稳定性（用氧化开始时的温度表征）和氧化反应动力学。可将已使用过的脂肪与新鲜脂肪区别开来。

The amorphous γ-form, also called the glass form, is obtained by shock-cooling the molten fat. Annealing causes this state to change to the α-and then finally to the stable β-configuration. Often, the goal of the industrial crystallization is to obtain the highest possible fraction of the stable high-melting β-modification.

The β'-form is relatively unstable, and sometimes a direct α-β-rearrangement occurs. The rearrangement processes are exothermic, whereas the melting is of course endothermic. Rearrangements involving unstable intermediates are readily detectable at low heating rates. As the crystallization temperatures can not be freely selected in the usual measurement methods, the obvious solution is to use the isothermic DSC operation as a routine method (cooling down to a crystallization temperature and keeping the temperature constant during the crystallization process).

The autoxidation of edible oils and fats can have a negative influence on the storage and processing of products which contain fat. At normal room temperatures, the oxidation is relatively slow, but the reaction rate increases rapidly at elevated temperatures in the presence of oxygen. Above 150 ℃, for instance in roasting or frying, autoxidation sets in and at higher temperatures leads even to complete decomposition of the fat.

The stability of the fat (characterized by the temperature at which the oxidation starts) and the reaction kinetics of the oxidation can be determined by DSC measurements. Used and fresh fats can be distinguished.

DSC 对脂肪的更多应用：
- 结晶行为的测试；在工业生产结晶过程工艺参数的影响；影响的来源
- 可可豆来源的测量
- 牛奶和乳脂中的脂肪酸组成和甘油三酯结构
- 巧克力的质量控制以及与感官品质的关系
- 固体脂肪指数（固体含量、液体含量）的测量
- 不同贮存条件下烹饪脂肪的测定
- 乳化剂（磷脂）对脂肪晶体结构的影响
- 棕榈油、棉花籽油、大豆油和许多其他纯甘油三酯的熔融和结晶行为

Additional Applications of DSC with Fats：
- Measurement of the crystallization behavior; determination of the influence of process parameters in the industrial crystallization; influence of the origin
- Determination of the origin of cocoa bean
- Fatty acid composition and structure of the triglycerides in milk and butter fats
- Quality control and correlation with sensory properties in the case of chocolate
- Determination of the solid fat index (solid fraction, liquid fraction)
- Behavior of cooking fat under different storage conditions
- Influence of emulsifying agents (phospholipids) on fat crystal structures
- Melting and crystallization behavior of pure trigylcerides such as palm oil, cotton seed oil, soybean oil and many others.

4.2.6　食品包装材料——塑料薄膜　Food Packagings-Plastic Films

塑料薄膜约为包装食品重量的百分之十。合适特定应用的薄膜复合物的种类是相当多的。作为进货检查和第三方产品监测与测定，薄膜分析是必不可少的。用 DSC 有可能鉴别复合物成分和辅料（例如黏合剂）。

可将试样放在敞口坩埚中，不做任何其他样品制备，以氮气气氛记录熔融曲线。由熔融温度可得到与聚合物性质有关的信息，由熔融焓还可获得与聚合物改性有关的信息。DSC 能解决许多问题，作为检查薄膜的快速方法很有用。对于更为广泛的表征，必定获得更多的信息，例如由溶解性测试、由切片薄片分析、由红外光谱或其他技术。无论何种情况，均可用 DSC 得到有价值的基础信息。

包装薄膜的质量控制

如果有参比材料，也能进行薄膜复合物的质量控制。直接用 DSC，而勿需样品的进一步制备，在几分钟内就能测定各种组分以及聚合物部分的不同构形。

Plastic films make up approximately ten percent by weight of packaged food products. The variety of film composites tailored for particular applications is enormous. For the checking of incoming goods and the monitoring and examination of third-party products an analysis of the film is necessary. Identification of the composite components and the auxiliaries (e. g. adhesives) is possible with DSC.

The melting curve is recorded without any further sample preparation in open pans under a nitrogen atmosphere. The melting temperatures provide information on the nature of the polymer. The enthalpies of fusion can also provide information on the modification of the polymer. DSC can answer many questions and is useful as a rapid method for checking films. For a more comprehensive characterization, additional information must be gained such as from solubility tests, from the examination of a microtome section, from an infrared spectrum or from other techniques. In all cases, valuable basic information can be obtained with DSC.

Quality Control of Packaging films

Quality control of film compostion can also be performed if reference material is supplied.

Various compositions and also different conformations of polymer fractions can be determined within minutes by DSC without further sample preparation.

用于表征薄膜的专门方法是固/液熔融相转变。这项 DSC 应用是很容易的。纯组分的熔点为确定温度,各组分的熔融焓比 DSC 噪声信号大若干数量级,这可保证极好的可再现熔融曲线。此外,在解释曲线时几乎无须考虑在有关的温度范围内发生副反应或其他相互作用。

对薄膜的其他应用:
- 熔点和结晶部分(结晶度)的测定

- 热历史对熔融行为的影响
- PE 密度对熔融行为和结晶的影响
- 结晶速率的估算
- 化学纯度的影响
- 熔融和氧化过程
- 聚合物等温氧化"诱导期"的测定

- 抗氧剂有效性的测试
- 半结晶聚合物无定形部分玻璃化转变温度和晶体部分熔点的测定

- 半结晶聚合物玻璃化转变温度、放热重结晶和吸热熔融的测定,例如 PET
- 用 TMA 测试玻璃化转变和熔融行为

The specific process that is used to characterize the film is the solid/liquid phase transition of melting. This is is a very easy DSC application. The melting points of the pure components are at definite temperatures and the enthalpy of fusion of the components is orders of magnitude greater than the DSC noise signal. This ensures perfectly reproducible melting curves. Furthermore, almost no side reactions or other interactions that must be taken into consideration in the interpretation of the curves, occur in the relevant temperature range.

Additional Applications with Films:
- Determination of the melting point and the crystalline fraction (crystallinity)
- Influence of the thermic history on the melting behavior
- Influence of the density of PE on the melting behavior and the crystallization
- Estimation of the crystallization rate
- Influence of the chemical purity
- Melting and oxidation processes
- Determination of the 'induction times' in the isothermal oxidation of polymers
- Measurement of the effectiveness of antioxidants
- Determination of the glass transition temperature of the amorphous fraction and the melting point of the crystalline fraction of semicrystalline polymers
- Determination of the glass transition temperature, exothermic recrystallization and endothermic melting of a semicrystalline polymer, e.g. PET
- Measurement of glass transitions and melting behavior by TMA.

5 热分析食品的典型应用
Typical TA Applications of Food

5.1 植物蛋白质的变性 Denaturation of Vegetable Proteins

样品 **Sample**	从有机食品商店购买的 4 种商品种子：小麦、斯佩耳特小麦、羽扇豆和大豆 Four commercial seeds from health food store: Wheat, Spelt, Lupines and Soybean	
条件 **Conditions**	测试仪器：DSC 坩埚：40 μl 铝坩埚，密封。 参比：密封坩埚中大约等量的水。	**Measuring cell**: DSC **Pan**: Al 40 μl, hermetically sealed. **Reference**: Approximately the same amount of water in a sealed pan.
	样品制备： 磨碎种子、在 pH 4.5 水相萃取、离心、除去浮层、在 pH 8.5 再萃取残留物、在真空下蒸发浮层至需要的干固含物。 测试： 以 5 K/min 由 30 ℃升温至 110 ℃	**Sample preparation**: Mill seeds, aqueous extraction at pH 4.5, centrifuge, discard supernatant, re-extract residue at pH 8.5, supernatant evaporated to desired dry substance content under vacuum. **Measurement**: Heating from 30 ℃ to 110 ℃ at 5 K/min

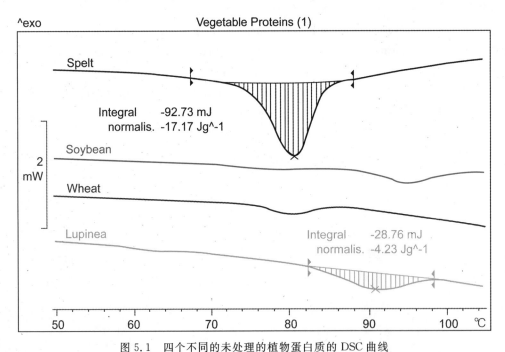

图 5.1 四个不同的未处理的植物蛋白质的 DSC 曲线
Figure 5.1 DSC curves of four different untreated vegetable proteins

解释 如图 4.1 所示，测试了四个不同的未经处理的植物蛋白质。每条曲线都在 72 ℃至 100 ℃范围呈现一个表征蛋白质变性的吸热峰。

Interpretation Four different untreated vegetable proteins were investigated as shown in figure 5.1. Each curve shows an endothermic peak in the range between 72 ℃ and 100 ℃ that is characteristic for protein denaturation. The proteins differ in

各蛋白质的热稳定性和变性焓是不同的。

对于具有可比的试样量和蛋白质含量,峰的积分为比反应焓的量度。

斯佩耳特小麦(Spelt)含有对温度最敏感的蛋白质,变性焓最大。小麦(wheat)蛋白质在86 ℃已深度变性,而大豆(soybean)蛋白质的主要部分仍然是原生态的,本例中,直到88 ℃才开始变性。

谷物和油料蛋白质的变性焓通常为3~10 J/g 蛋白。焓变取决于蛋白质相,因而原始状态的种子蛋白质的变性焓比水溶液中溶解状态的要小。

天然蛋白质变性焓越大,就越有利于对经处理的蛋白质试样中的变性蛋白质部分进行DSC测量。

DSC可自由选择升温速率,这有助于改善对反应的测试和检测。升温速率越快,测得的信号就越大,即测得的峰越大。当然,如果反应过程不变,算得的峰面积是相同的。不过,应注意到的是,快的升温速率使反应移至较高的温度。

DSC鉴定植物蛋白质在实践中应用较少,即使应用也会很复杂,因为反应焓随蛋白质种类及其生长条件而异。因此,当估算变性程度时,须将天然蛋白质作为参比进行测试。

DSC曲线是样品测定的"指纹"。食品中蛋白质部分的这种指纹图是蛋白质品质的特征。

their thermic stability and enthalpy of denaturation.

With comparable sample weights and protein concentrations, the integrals of the peaks are a direct measure of the specific reaction enthalpies.

Spelt contains the proteins that are most temperature sensitive and that have the highest enthalpy of denaturation. Whereas the wheat proteins are already extensively denatured at 86 ℃, the main fraction of the soybean protein is still native. In this case, denaturation does not set in until 88 ℃.

The enthalpies of denaturation for grain and oil seed proteins usually lie in the range 3~10 J/g protein. The enthalpy changes depend on the phase of the proteins, so that the enthalpy of denaturation of seed proteins in the original condition is less than in the dissolved state in aqueous solutions.

The higher the enthalpy of denaturation of the native protein, the better the measurement of the fractions of denatured protein in processed protein samples by DSC.

The free selection of the heating rate with DSC can be of use to help improve the measurement and detection of reactions. The higher the heating rate, the larger the detected signal, i. e. the larger the measured peak. The calculated peak area should of course be the same if the reaction process is the same. However, it should be noted that at high heating rates the reaction is shifted to higher temperatures.

Identification of vegetable proteins by DSC is an application that is less likely to be used in practice and would also be complicated since the reaction enthalpy can vary depending on the variety and growth conditions. A native protein must, therefore, always be measured as a reference when estimating the degree of denaturation.

The DSC curve is a 'fingerprint' of the measured sample. Such fingerprints of protein fractions in foods are characteristic of the condition of the protein.

计算 Evaluation	样品 Sample	DS 含量 DS content % DS	试样量 Sample weight mg DS	反应焓 Reaction enthalpy J/g
	小麦 Wheat	12.5	3.44	3.5
	斯佩耳特小麦 Spelt	20.0	5.40	17.2
	羽扇豆 Lupines	23.0	6.80	4.2
	大豆 Soybean	13.0	4.78	5.2

DS=干固物
DS=dry substance

结论. 所讨论的测试表明,原则上,含蛋白质的任何产品都可用DSC进行研究。更可用于动物蛋白质(参见鸡蛋、血和肌肉蛋白)的测试,因为通常在动物产品(肌肉)中原始蛋白质含量越高,则测试灵敏度就越高。并可检测其存在于水的环境。

Conclusion The measurements discussed show that, in principle, any product which contains protein can be investigated by DSC. This applies all the more to animal proteins (see also egg, blood and muscle protein) owing to the generally higher original protein concentration in animal products (muscle), the greater measurement sensitivity and their occurrence in a aqueous environment.

5.2 鸡蛋蛋白质的变性　Egg Protein Denaturation

样品　鸡蛋蛋清
Sample　Egg white of a hen's egg

条件　测试仪器:DSC
Conditions
坩埚:40 μl 铝坩埚,密封。
参比:大约等量的水。
样品制备:
将蛋清分离,搅拌 2 min。干固物含量:DS=11.6%。试样质量 32.9 mg。
测试:
以 10 K/min 由 30 ℃升温至 110 ℃

Measuring cell:DSC
Pan:Al 40 μl, hermetically sealed.
Reference:Approximately the same amount of water.
Sample preparation:
Egg white was separated and stirred for 2 min. Content of dry substance:DS=11.6%. Sample mass 32.9 mg.
Measurement:
Heating from 30 ℃ to 110 ℃ at 10 K/min

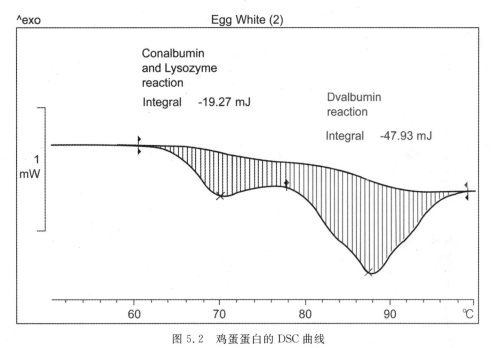

图 5.2 鸡蛋蛋白的 DSC 曲线
Figure 5.2 DSC curve of Egg white of a hen's egg

解释 如图 5.2 所示,将新鲜鸡蛋蛋清升温,在 70 ℃ 和 87 ℃ 呈现两个吸热主峰(双峰)。第 1 峰与伴清蛋白部分(总蛋白质的 13.8%,变性热为 19.8 J/g[1])的变性有关,第 2 峰代表卵清蛋白部分(总蛋白质的 65%,变性热为 28.9 J/g[1])的变性。由于溶菌酶部分(总蛋白质的 3.4%[1])的变性介于两峰之间,所以第 1 峰不能与第 2 峰完全分开。

Interpretation As shown in figure 5.2, heating up the egg white of a fresh hen's egg shows two main endothermic peaks at 70 ℃ and at 87 ℃ (double peak). The first peak relates to the denaturation of the conalbumin fraction (13.8% of the total protein, heat of denaturation 19.8 J/g [1]), the second peak represents the denaturation of the ovalbumin fraction (65% of the total protein, heat of denaturation 28.9 J/g[1]). Since the lysozyme fraction (3.4% of the total protein [1]) denatures between the two main peaks, the the first peak is not completely separated from the second peak.

计算
Evaluation

	第 1 峰 First peak	第 2 峰 Second peak
吸热面积 Endothermic area mJ	19.3	47.9
峰温 Temperature of peak ℃	69.8	87.4

结论 典型的变性峰可用于鉴定。

Conclusion The typical denaturation peaks can be used for identification purposes.

[1] Donovan, J. W. et al.: A differential scanning calorimetric study of the stability of egg white to heat denaturation, J. Sci. Fd. Agric. 26:73, 1975

5.3 鸡蛋蛋清热处理的影响
Influence of Thermal Treatment of Egg White

样品 **Sample**	鸡蛋蛋清 Egg white of a hen's egg		
条件 **Conditions**	测试仪器：DSC 坩埚：40 µl 铝坩埚，密封。 参比：大约等量的水。	**Measuring cell**：DSC **Pan**：Al 40 µl, hermetically sealed. **Reference**：Approximately the same amount of water.	

样品制备：
蛋清称量后，将密封的坩埚浸入热水浴(80 ℃)中一定时间。

Sample preparation：
After weighing the egg white, the sealed pan is immersed in a thermostatted water bath (80 ℃) for the specified exposure times.

测试：
以 10 K/min 由 30 ℃升温至 110 ℃

Measurement：
Heating from 30 ℃ to 110 ℃ at 10 K/min

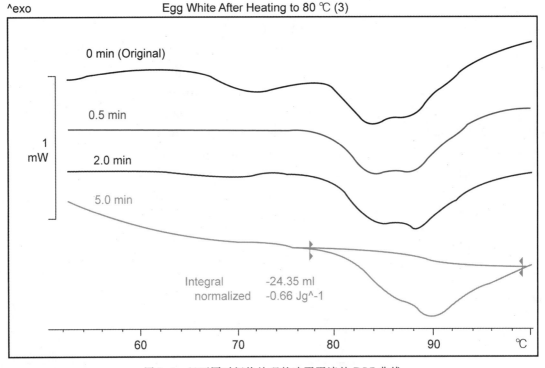

图 5.3 经不同时间热处理的鸡蛋蛋清的 DSC 曲线
Figure 5.3 DSC curves of thermal treated egg white for different time

解释 如图 5.3 所示，在预处理期间发生变性，因而残余反应焓随着浸浴时间增长而下降。经 30 s 处理的试样已检测不到伴清蛋白部分，该试样加热直至 80 ℃仍无反应峰。83 ℃和 88 ℃处的双峰为完好卵清

Interpretation As shown in figure 5.3, denaturation occurs during the pretreatment so that the remaining reaction enthalpy decreases with increasing exposure. The conalbumin fraction is no longer detected in the sample treated for 30 seconds, which shows there is no reaction peak until 80 ℃ which the sample has been heated to. The peak with a double maximum at 83 ℃ and

蛋白的变性峰。随着处理时间增长,该峰向更高温度移动,经 5 min 处理的试样,峰值移至 90 ℃。卵清蛋白经该热引发重排转变为热更稳定的形式 S-卵清蛋白(参见下一应用实例)。反应焓保持不变的事实表明,卵清蛋白和 S-卵清蛋白经热处理均无明显受损。

经由中间产物发生重排,是不可逆的。耐热性提高是共价键结构改变的结果。pH 值从 8.1 提高到 9.0,转变为较耐热结构的量会使通常的变性焓大约改变 10%。

88 ℃ shows the intact ovalbumin denaturation peak. With increasing treatment time, the peak shifts to higher temperature and to an absolute maximum at 90 ℃ with the sample treated for 5 minutes. This thermicly induced rearrangement is the transition from ovalbumin to the thermicly more stable form S-ovalbumin (see also the next application example). The fact that the reaction enthalpy remains constant shows that neither the ovalbumin nor the S-ovalbumin fraction are significantly damaged by the heat treatment.

The rearrangement is irreversible and takes place via intermediate products. The increased thermic resistance is a consequence of a change in the covalent structure. The pH value increases from 8.1 to 9.0. The transition to the more heat resistant structure amounts only to about 10% of the usual enthalpy of denaturation.

计算
Evaluation

浸浴时间 Exposure time min	试样量 Sample weight mg	反应焓 Reaction enthalpy J/g(蛋清 egg white)
0.0	25.03	1.8
0.5	25.14	1.0
2.0	29.32	0.8
5.0	36.64	0.7

结论 测试表明,DSC 既可确定热处理的程度(验证热处理),又可测定过程参数温度和时间对蛋白质质量的影响(用于过程优化)。

Conclusion The measurements show that DSC can both verify the extent of thermal treatment (for proof of the treatment) and determine the influence of the process parameters temperature/time on the protein quality for process optimization purposes.

5.4 鸡蛋贮存时间的影响　Influence of Egg Storage Time

样品　市售鸡蛋,A 级
　　　　包装时间:6 月 6 日
　　　　包装时间:6 月 18 日

Sample　Commercially available eggs, class A
　　　　Packaging date: June 6
　　　　Purchase date: June 18

条件　测试仪器:DSC
Conditions　坩埚:40 μl 铝坩埚,密封。
　　　　参比:大约等量的水。
　　　　样品制备:
　　　　在 30 ℃的生理盐水溶液中贮存。贮存期结束时,将蛋清分离,搅拌 2 min。

Measuring cell: DSC
Pan: Al 40 μl, hermetically sealed.
Reference: Approximately the same amount of water.
Sample preparation:
Storage of the eggs in a physiological saline solution at 30 ℃. At the end of the storage time, the egg white was separated and stirred for two minutes.

测试：以 5 K/min 由 30 ℃升温至 110 ℃

Measurement: Heating from 30 ℃ to 110 ℃ at 5 K/min

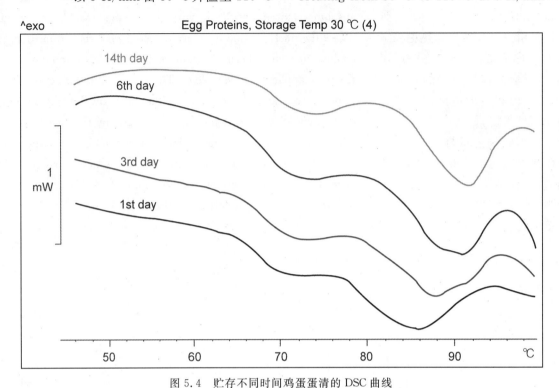

图 5.4　贮存不同时间鸡蛋蛋清的 DSC 曲线

Figure 5.4　DSC curves of egg white stored for different time

解释　在食物内，除了由于热或 pH 的影响引起的自发转变，还会发生缓慢的构形变化，如果所贮存的产品的反应行为发生改变，则可用 DSC 进行测定。例如，需要的话，可推断出贮存时间（还包括肉的保质期）。

上述反应包括蛋清中卵清蛋白转变为 S-卵清蛋白的蛋白重排（或碳水化合物情况下的凝沉）。如图 5.4 所示，新鲜蛋清在 86 ℃呈现卵清蛋白的变性峰（第 1 天）。贮存仅 2 天时间后，在 88 ℃和 91 ℃出现两个最大值，表明转变为 S-卵清蛋白的重排已经开始。5 天后，主最大值已移至 91 ℃。14 天后，全部蛋白部分已变为 S-构形，峰最大值在 92 ℃。峰温的移动与贮存时间直接相关。

Interpretation　In addition to the spontaneous transitions due to the influence of heat or pH, slower conformational changes also occur in food and can be measured by DSC if the reaction behavior of the stored products changes. For instance, the storage time (including also the post-mortem phase of meat) can be deduced if required.

The above-mentioned reactions include the protein rearrangement of ovalbumin in egg white to S-ovalbumin (or retrogradation in the case of carbohydrates). As shown in figure 5.4, fresh egg white shows an ovalbumin denaturation peak at 86 ℃ (first day). After a storage time of just two days, two maxima appear at 88 ℃ and 91 ℃, a sign that the rearrangement to S-ovalbumin has started. After five days the main maximum has shifted to 91 ℃. After 14 days the entire protein fraction has changed to the S-conformation with a peak maximum at 92 ℃. The shift in the peak temperature correlates directly with the storage time.

计算
Evaluation

日期 Date	试样量 Sample weight mg	试样 Sample	峰1 Peak 1 ℃	峰2 Peak 2 ℃
6月19日 June 19	27.38	第1天 1st day	70.4	86.2
6月21日 June 21	27.38	第3天 3rd day	72.1	88.4
6月24日 June 24	33.66	第6天 6th day	72.9	91.0
7月2日 July 02	30.47	第14天 14th day	74.0	92.0

结论 即使新鲜样品与贮存14天的样品间的峰最大值仅移动不到6 K，但测量重复性很好，因而在贮存的最初10天内，可方便地将贮存时间估算到最近的两天。在卵清蛋白部分完全重排转换为S-卵清蛋白后，则无法跟踪贮存时间。

Conclusion Even though the shift of the peak maximum between the fresh sample and the sample stored for 14 days is only less than 6 K, the measurement is so reproducible, that within the first ten days of storage the storage time can be readily estimated to the nearest two days. After complete rearrangement of the ovalbumin fraction to S-ovalbumin, it is no longer possible to follow the storage time.

5.5 pH对牛血红蛋白的影响
Influence of pH on Bovine Hemoglobin

样品 牛全血，用柠檬酸盐稳定
Sample Bovine whole blood, stabilized with citrate

条件 测试仪器：DSC
Conditions 坩埚：40 μl 铝坩埚，密封。
参比：大约等量的水。

Measuring cell：DSC
Pan：Al 40 μl, hermetically sealed.
Reference：
Approximately the same amount of water.

样品制备：
通过冷冻破坏细胞。原始pH为7.06，用0.5 mol/L HCl进行调节。

Sample preparation：
Destroying the cells by freezing. The native pH is 7.06, the adjustment has been done with 0.5 mol/L HCl.

测试：
以10 K/min由30 ℃升温至110 ℃

Measurement：
Heating from 30 ℃ to 110 ℃ at 10 K/min

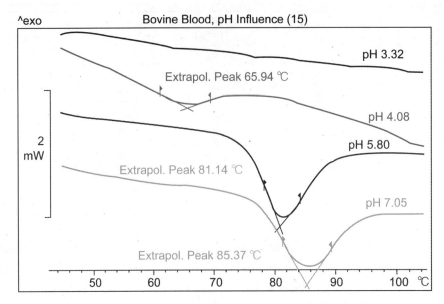

图 5.5　不同 pH 值下红血球的 DSC 曲线
Figure 5.5　DSC curves of red blood corpuscles exposed to different pH values

解释　实际 pH 值及其是否保持不变或变化,对于加工过程和产品质量具有很大影响。

图 5.5 为处于不同 pH 值时红血球即血红蛋白主要蛋白质成分的 DSC 曲线。血红蛋白链,即血红蛋白的一种蛋白质部分,加热时变性。

血的生理 pH 值为 7.35。此时,无法测量血红蛋白,因为它会立即凝结。在 pH 值 7.06 添加少量柠檬酸进行测量,测得值 18 J/g 作为原始的变性焓。

当将 pH 稍降至酸性值(pH 5.8)时,变性焓没有明显下降,但峰最大值移至较低的温度。换言之,pH 值影响蛋白的热稳定性。

原血红素使血红蛋白的结构稳定。在酸性变性中,分子变得不稳定,从而对热较敏感。一个显著特征是峰最大值的降低,而且峰变宽。在 pH 4.28,分子伸展,但原血红素仍然附着在血红蛋白链上。只可测得很小的焓变,反应发生在较低的温度(60 ℃至 70 ℃)。

在酸性 pH 范围(pH 3.3),样品已经完全变性,即由于氢离子浓度高而解开缠结,因而 DSC 测试的热处

Interpretation　The actual pH value and whether it is kept constant or changes has a large influence on the course of the process and the product quality.

Figure 5.5 shows the DSC curves of the main protein component of the red blood corpuscles, hemoglobin, exposed to different pH values. The globin chains, which are a protein fraction of the hemoglobin molecule, are denatured on heating.

The physiological pH value of the blood is 7.35. In this state, the hemoglobin can not be measured because it immediately coagulates. A value of 18 J/g is assumed to be the native denaturation enthalpy measured with addition of a small amount of citrate at pH 7.06.

When the pH is lowered to a slightly acidic value (pH 5.8), the enthalpy shows no significant decrease, but the peak maximum is shifted to lower temperatures. In other words the pH value influences the thermic stability of the protein.

Heme stabilizes the structure of hemoglobin. In an acidic denaturation, the molecule becomes unstable and hence more sensitive to heat. A significant feature is the lowering of the temperature of the peak maximum. Moreover the peaks become broader. At pH 4.28 the molecule unfolds, but the heme is still attached to the globin chains. Only a very small enthalpy change can be measured and the reaction takes place at much lower temperatures (60 ℃ to 70 ℃).

In the acidic pH range (pH 3.3), the sample is already fully denatured, i. e. uncoiled as a result of the high hydrogen ion concentration, so that the thermic treatment of the DSC measurement

理不会引起进一步反应,记录得到的只是一条直线基线。

induces no further reaction. Only a straight baseline is recorded.

计算
Evaluation

试样 pH Sample pH	试样量 Sample weight mg	峰温 Peak temperature ℃
7.06	29.51	85.4
5.80	29.07	81.1
4.08	29.12	65.9
3.32	22.28	—

结论 用 DSC 可方便地测定蛋白质的稳定性和加工条件的影响。

Conclusion The stability of the proteins and the effect of the process conditions are readily shown by DSC.

5.6 肉类的 DSC　DSC of Meat

样品　鸡肉、火鸡肉、小牛肉
Sample　Chicken, turkey and veal

条件　测试仪器:DSC
Conditions　坩埚:40 μl 铝坩埚,密封。
参比:
大约 70 mg 氧化铝,以补偿试样的热容。
样品制备:
用锐利的刀将样品切成碎片。
测试:
以 10 K/min 由 30 ℃升温至 110 ℃

Measuring cell:DSC
Pan:Al 40 μl, hermetically sealed.
Reference:
Approximately 70 mg alumina to compensate the heat capacity of the sample.
Sample preparation:
Small pieces were cut with a sharp knife.
Measurement:
Heating from 30 ℃ to 110 ℃ at 10 K/min

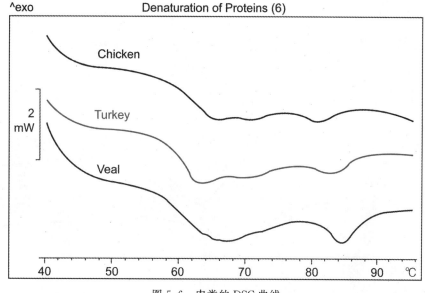

图 5.6　肉类的 DSC 曲线
Figure 5.6　DSC curves of meat

解释 如图 5.6 所示，肉蛋白质经历若干个构象的变化。在 55 ℃至 62 ℃肌球蛋白及其中间产物发生转变，在约 67 ℃肌质蛋白和胶原蛋白发生转变，在 78 ℃ 和 83 ℃ 肌动蛋白发生转变[1]。DSC 曲线是各种肌肉蛋白的特征"指纹"。

结论 DSC 曲线显示各个变性温度，它们是加工的关键参数，可用于鉴别。

Interpretation The meat proteins undergo several conformational changes, as displayed in figure 5.6. There are the transitions of myosin and sub-units at 55 ℃ to 62 ℃, of sarcoplasmic proteins and collagens around 67 ℃ and of actin at 78 ℃ to 83 ℃ [1]. The DSC curves are characteristic 'fingerprints' of the various muscle proteins.

Conclusion The DSC curves show the denaturation temperatures. They are the key parameters for processing and serve for identification purposes.

[1] Harwalkar and Ma, Thermic Analysis of Foods, Elsevier, 1990, p 93.

5.7 淀粉的凝胶化　Gelatinization of Starch

样品 **Sample**	玉米淀粉、大米淀粉、小麦淀粉、马铃薯淀粉 Corn starch, rice starch, wheat starch, potato starch	
条件 **Conditions**	测试仪器：DSC 坩埚：40 μl 铝坩埚，密封。 样品制备： 通过搅拌制备淀粉水悬浮液（20%水），将均匀的悬浮液放入坩埚称重。 测试： 以 10 K/min 由 30 ℃升温至 110 ℃	**Measuring cell**：DSC **Pan**：Al 40 μl, hermetically sealed. **Sample preparation**: Preparation of a suspension of starch in water (20 weight%) by stirring, homogenized suspension weighed into the pan. **Measurement**: Heating from 30 ℃ to 110 ℃ at 10 K/min

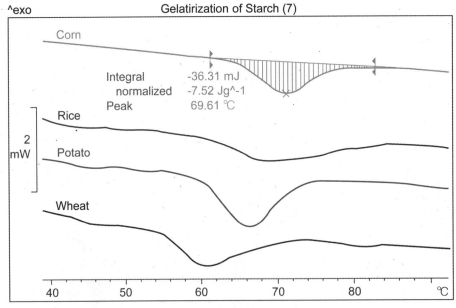

图 5.7 不同淀粉的 DSC 曲线

Figure 5.7 DSC curves of different starch

解释 图 5.7 所示 DSC 曲线较大的波动产生于试样的不均匀性,即不同的淀粉块随机发生反应。但淀粉溶胀峰是容易分辨的。

Interpretation The relatively high fluctuations in the DSC curve displayed in figure 5.7 can originate from the inhomogeneity of the sample, i.e. various starch clusters react stochastically. However, the peak of the starch swelling is easily recognized.

计算
Evaluation

试样 Sample	试样量 Sample weight mg	反应焓 Reaction enthalpy J/g	峰温 Peak ℃
玉米 Corn	4.83	7.5	69.6
大米 Rice	5.18	8.8	66.3
马铃薯 Potato	5.33	11.0	64.4
小麦 Wheat	5.46	8.1	59.2

结论 DSC 曲线显示凝胶化温度,是加工的关键参数,可用于鉴定。

Conclusion The DSC curves show the gelatinization temperatures. They are key parameters for processing and serve to identify the different starches.

5.8 水中淀粉含量对溶胀的影响
Influence of the Starch Content on Swelling in Water

样品　马铃薯淀粉
Sample　Potato starch

条件　测试仪器:DSC。
Conditions
坩埚:40 μl 铝坩埚,密封。
样品制备:
通过加入不同量蒸馏水制备淀粉悬浮液;搅拌(用 Ultra-Turrax 搅拌 1 min);将均匀的悬浮液称入坩埚。

测试:
以 5 K/min 由 20 ℃升温至 110 ℃

Measuring cell: DSC
Pan: Al 40 μl, hermetically sealed.
Sample preparation:
Preparation of the starch suspensions by the addition of varying amounts of distilled water; stirring (Ultra-Turrax for 1 min); homogenized suspension weighed into the pan.

Measurement:
Heating from 20 ℃ to 110 ℃ at 5 K/min

解释 淀粉溶胀的决定性因素之一是淀粉/水的比例。
水过量(大于每脱水葡萄糖单元4个水分子)时,吸热峰只有约10 K宽,参见上例 20% 淀粉水溶液的 DSC 曲线(图5.7)。如果淀粉/水之比升至 40% 以上,则反应温度范围变宽,

Interpretation One of the most decisive factors in the swelling of starch is the starch/water ratio.
With an excess of water (more than 4 molecules of water per anhydroglucose unit), the endothermic peak is only about 10 K wide, which is shown by the previous example with 20% starch in water (figure 5.7). If the starch/water ratio rises above a value of 40%, the reaction temperature range broadens or a

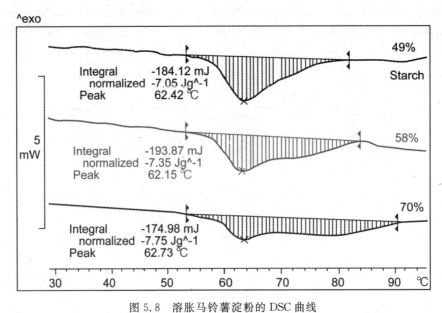

图 5.8 溶胀马铃薯淀粉的 DSC 曲线
Figure 5.8　DSC curves of swelled potato starch

或形成第 2 吸热峰,如图 5.8 所示。60 ℃至 80 ℃间的第 1 峰出现在恒定的温度,峰温约为 62 ℃,而在更高温度,显然发生了反应,开始时只在第 1 峰形成一个肩,反应的温度范围模糊不清,但在更高比例时,出现第 2 峰。第 1 峰的面积下降,但总焓保持不变。

second endothermic peak is formed, as shown in figure 5.8. Whereas the first peak between 60 ℃ and 80 ℃ appears at constant temperature with a peak at about 62 ℃, a reaction at higher temperatures is evident, which initially leads only to the formation of a shoulder on the first peak and the blurring of the temperature range of the reaction but then, at higher ratios, to the appearance second peak. The area of the first peak decreases, but the total enthalpy remains the same.

在上例测试的水有限过量(20%)的情况下,吸热峰介于 60 ℃至 70 ℃间,而淀粉为 70% 时整个反应超过 30 K 范围。

Whereas in the limiting case of excess water (20% starch) in the above measurement the endothermic peak lies between 60 ℃ and 70 ℃, the overall reaction at 70% starch extends over a range of 30 K.

水含量较高时,系统由于水合作用(吸水)而不稳定。聚合物链的加剧运动(吸热)和淀粉颗粒的熔化开始(第 1 吸热峰)。在更浓的淀粉溶液中,该效应由于水含量的限制而下降。淀粉颗粒直接的热熔化出现在较高温度。

At high water contents, the system is destabilized through hydration (water uptake). Increased motion of the polymer chains (heat uptake) and melting of the starch granules begins (first endotherm). In more concentrated starch solutions, this effect is reduced as a result of the limited water content. Direct thermic melting of the starch granules occurs at higher temperatures.

计算
Evaluation

淀粉含量 Starch content %	试样量 Sample weight mg	反应焓 Reaction enthalpy J/g	峰温 Peak ℃
49	26.12	7.0	62.4
58	26.37	7.4	62.2
70	22.58	7.8	62.7

结论 这些 DSC 曲线表明，水/淀粉比对于结果的再现性很重要。建议将淀粉和水直接称入坩埚（无需搅拌；扩散会调节局部浓度梯度）。

Conclusion These DSC curves show that the water/starch ratio is important for reproducible results. We recommend weighing starch and water directly into the pan (no mixing required; diffusion adjusts local gradients).

5.9 无定形糖的 ADSC(调制 DSC)　ADSC of Amorphous Sugar

样品	糖(蔗糖)	
Sample	Sugar（Saccharose）	
条件	测试仪器：DSC	**Measuring cell**：DSC
Conditions	坩埚：40 μl 铝坩埚，盖钻孔。	**Pan**：Al 40 μl, with pierced lid.

样品制备：

将糖放入坩埚称重，以 5 K/min 在 DSC 样品皿中加热。在吸热峰后立即移走坩埚，放于铝板上骤冷至室温，以制取呈非晶相的样品。

Sample preparation:

The sugar is weighed in the pan and heated in the DSC cell at 5 K/min. Immediately after the endothermic fusion peak the pan is removed and shock-cooled to ambient temperature on an aluminum plate in order to obtain the amorphous phase.

测试：

以平均 2 K/min 的升温速率由 20 ℃ 升温至 110 ℃。调制振幅 1 ℃，周期 1 min。在相同条件下对空样品皿和铝试样进行测量，用作校准。

Measurement:

Heating from 50 ℃ to 170 ℃ at a mean heating rate of 2 K/min. Modulation amplitude 1 ℃, period one minute. The empty cell and an aluminum sample are measured under the same conditions for calibration purposes.

气氛：氮气，50 ml/min

Atmosphere：Nitrogen, 50 ml/min

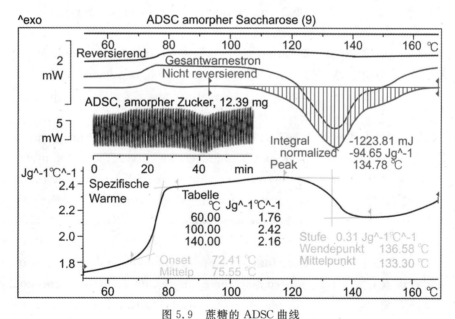

图 5.9　蔗糖的 ADSC 曲线

Figure 5.9　ADSC curves of Saccharose

解释 在图 5.9 的中间坐标系中，所示曲线为用调制温度程序得到的

Interpretation In the middle coordinate system of figure 5.9, the original DSC curve obtained using the alternating temperature

原始 DSC 曲线,呈现许多的变化。快速傅里叶分析能获得平均信号值(=总热流信号,等于以 2 K/min 测量的常规 DSC 曲线)、振幅曲线(可逆曲线)以及所谓的不可逆曲线(=总热流曲线-可逆曲线)。

在可逆曲线上,可观察到由于玻璃化转变、结晶和其他现象产生的比热容变化。由可逆曲线还可计算表示 c_p 变化的比热容与温度的关系(图中下面曲线),直接以 J/gK 表示。

不可逆效应包括化学反应、加热无定形样品时发生的冷结晶、玻璃化转变时的松弛效应和挥发物的蒸发(干燥)等。

有很多通常结晶的物质,当从熔融态骤冷时,形成玻璃态即非晶聚集态。这种物质的例子有硫磺、多数有机化合物、一些半结晶聚合物、某些合金和石英。加热时,经历玻璃化转变变软成为橡胶态。由于分子运动,然后能形成晶体(去玻璃化),该结晶过程在 DSC 曲线(总热流和不可逆曲线)上产生放热峰。

计算 75.6 ℃ 的玻璃化转变中点对于无水糖是典型的;水分会起增塑剂作用,使玻璃化转变温度范围降低。

由不可逆曲线上结晶峰的积分得到结晶热。94.6 J/g 的值与完全结晶并不相等,因为完全结晶的糖的熔融热约为 130 J/g。对应的 c_p 变化为 0.31 J/gK。

结论 在食品加工中,玻璃化转变温度与糖果的稳定性因素有关、与其他非结晶干燥食品有关、与速冻过程有关,已变得越来越重要。

温度调制 DSC 最引人注意的特点是能将 c_p 从一些如结晶所示的焓变中分离开来。

program is displayed. It shows a number of changes. Fast Fourier Analysis allows the recording of the mean signal (= total heat flow signal, corresponding to a classical DSC curve measured at 2 K/min), the amplitude curve (reversing curve) and the so called non-reversing curve (= total heat flow curve-reversing curve).

The changes of the specific heat capacity due to the glass transition, crystallization and other phenomena can be observed in the reversing curve. The reversing curve allows the calculation of the temperature function of the specific heat capacity (lower diagram) again showing the c_p changes but in this case directly in J/gK.

The non-reversing effects include for example chemical reactions, cold crystallization on heating amorphous samples, relaxation effects during the glass transition and the evaporation of volatiles (drying).

There is a broad range of substances that are usually crystalline but, when shock-cooled from the molten phase, form a glassy or amorphous state of aggregation. Examples of such substances are sulfur, most organic compounds, some semicrystalline polymers, certain alloys and quartz. On heating they undergo a glass transition and become soft and rubbery. Due to the mobility of the molecules, they are then able to form crystals (devitrification). This crystallization process leads to an exothermic peak on the DSC curve (total heat flow and non-reversing curve).

Evaluation The glass transition midpoint of 75.6 ℃ is typical for anhydrous sugar; moisture would act as a plasticizer and lower the temperature range of glass transition.

The heat of crystallization is obtained by integrating the crystallization peak on the nonreversing curve. The value of 94.6 J/g does not correspond to complete crystallization since the heat of fusion of completely crystalline saccharose is approximately 130 J/g. The corresponding change in c_p is 0.31 J/gK.

Conclusion In food technology the glass transition temperature is becoming increasingly important in connection with stability considerations of sweets, with other amorphous dried foods and with deep-freeze processes.

The most attractive feature of temperature-modulated DSC is its ability to separate c_p changes from certain enthalpy changes as shown in the case of crystallization.

5.10 糖和淀粉的 TGA　TGA of Sugar and Starch

样品 **Sample**	糖(蔗糖)、玉米淀粉 Sugar (saccharose), corn starch		
应用 **Application**	生产乳膏的原料 Basic materials for the manufacture of creams		
条件 **Conditions**	测试仪器：TGA	**Measuring cell**：TGA	
	坩埚：70 μl 氧化铝坩埚,无盖	**Pan**：Alumina 70 μl, no lid	
	样品制备：	**Sample preparation**：	
	分别将 5.26 mg 糖和 4.99 mg 淀粉称入坩埚。如果以 10 K/min 将糖加热,则会在分解时起泡,溢出坩埚。加入约 20 mg 干燥氧化铝粉末有助于使试样保持在坩埚中。	Weigh 5.26 mg sugar and 4.99 mg starch into pans respectively. If sugar is heated at a rate of 10 K/min it foams up during decomposition and may leave the pan. The addition of approximately 20 mg dry alumina powder helps to keep the sample in the pan.	
	测试：	**Measurement**：	
	以 10 K/min 由 30 ℃升温至 600 ℃	Heating from 30 ℃ to 600 ℃ at 10 K/min	
	气氛：氮气,50 ml/min	**Atmosphere**：Nitrogen, 50 ml/min	

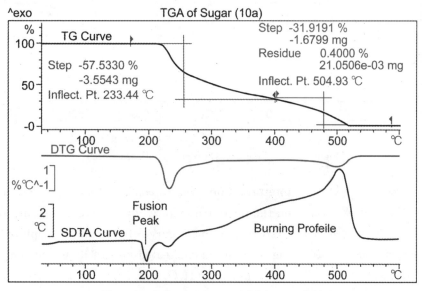

图 5.10　糖的 TGA、DTG 和同步 DTA 曲线

Figure 5.10　TGA, DTG and simultaneous DTA (SDTA) curves of sugar

解释　如图 5.10 糖的 TGA 测量曲线所示,高至 200 ℃的 TGA 曲线平坦部分证明糖中没有水分(<0.1%)。发生的第 1 个过程是在 190 ℃的熔融,只在同步 DTA(SDTA)曲线上可观察到。在液相,碳水化合物失去水并熔化变成焦糖。按化学计量,由分

Interpretation　As shown by the TGA measurement curve of sugar displayed in figure 5.10, the flat part of the TGA curve up to 200 ℃ proves that there is no moisture in the sugar (<0.1%). The first process that occurs is melting at 190 ℃, which is only visible in the SDTA curve. In the liquid phase the carbohydrate loses water and caramelizes. Stoichiometrically, from the formula $C_n(H_2O)n$ one expects the formation of 60%

子式 $C_n(H_2O)_n$ 可预计形成 60% 的水和 40% 的炭黑。但是，由于发生了其它反应，因而并没有清晰的失水台阶。

计算 DTG 最小值通常用作分开重叠台阶的计算界限。67.3% 的失水台阶接近于上述 60% 的值。生成的炭黑放热燃烧至 540 ℃。SDTA 曲线的形状称为"燃烧曲线"，表示炭黑的反应性。590 ℃ 处 0.04% 的残余物为矿物灰分

water and 40% carbon black. But, there is no distinct dehydratation step because of concurrent other reactions.

Evaluation The DTG minimum is normally used as the evaluation limit to separate overlapping steps. The dehydratation step of 67.3% is close to the above mentioned value of 60%. The carbon black formed burns exothermicly up to 540 ℃. The shape of the SDTA curve is called the 'burning profile' and gives an indication of the reactivity of the carbon black. The residue of 0.40% at 590 ℃ is the mineral ash content.

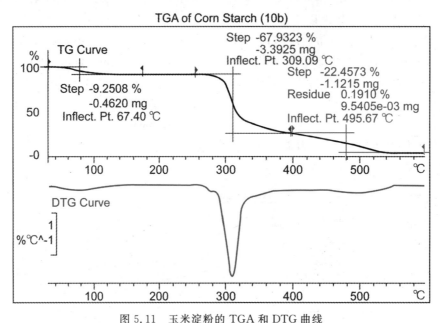

图 5.11　玉米淀粉的 TGA 和 DTG 曲线
Figure 5.11　TGA and DTG curves of corn starch

解释 取决于环境空气的相对湿度，淀粉含有若干百分点的水。如图 5.11 玉米淀粉的 TGA 测量曲线所示，失去水分高至 200 ℃。按化学计量，由分子式 $C_n(H_2O)_n$ 可预计形成 60% 的水和 40% 的炭黑。同样，由于发生了其它反应，没有清晰的失水台阶。

Interpretation Starch contains several percent moisture depending on the relative humidity of the surrounding air. As shown by the TGA measurement curve of corn starch displayed in figure 5.11, the moisture is eliminated up to 200 ℃. Stoichiometrically from the formula $C_n(H_2O)_n$ one expects 60% water and 40% carbon black. Again, there is no distinct dehydratation step because of concurrent other reactions.

计算 DTG 最小值通常用作分开重叠台阶的计算界限。检测到 9.2% 的水分。接下来 67.3% 的台阶高于 60% 的预计值。生成的炭黑放热燃烧至 540 ℃。590 ℃ 处 0.19% 的残余物为矿物灰分。

Evaluation The DTG minimum is used as the evaluation limit to separate overlapping steps. There is 9.2% of moisture detected. The next step of 67.9% is higher than the value expected of 60%. The carbon black formed burns up to 540 ℃. The residue of 0.19% at 590 ℃ corresponds to the mineral ash content.

结论 TGA 可测定水分含量、活性成分含量和灰分含量。此外,同步 DTA 曲线上的熔点和 TGA 拐点温度可用于鉴别不同的碳水化合物。

Conclusion TGA allows the determination of the moisture content, the content of active ingredients and the ash content. In addition, the melting point on the simultaneous DTA curve and the TGA inflection temperatures are used to identify the different carbohydrates.

5.11 意大利通心粉的动态负载 TMA
Dynamic Load TMA of Pasta

样品	面条,0.8 mm 厚
Sample	Noodle, 0.8 mm thick
条件	测试仪器:TMA
Conditions	探头:直径 1.1 mm

样品制备:
由面条折断得到约 2 mm×2 mm 的片(太大的试样在加热时可能因水分蒸发而发生爆炸)。样品已在约 50%相对湿度下放置 1 天。

测试:
以 10 K/min 由 30 ℃升温至 150 ℃
动态负载:0.05/0.20 N、周期 12 s
气氛:氮气,空气,静止环境,无流动

Measuring cell:TMA
Probe:1.1 mm diameter

Sample preparation:
A piece of approx. 2 mm×2 mm is broken fo tfhe noodle (a sample that is too large may explode on heating due to vaporization of moisture). The sample has been kept at approx. 50% relative humidity for 1 day.

Measurement:
Heating from 30 ℃ to 150 ℃ at 10 K/min
Dynamic load:0.05/0.20 N, period:12 seconds
Atmosphere:
Air, stationary environment, no flow rate

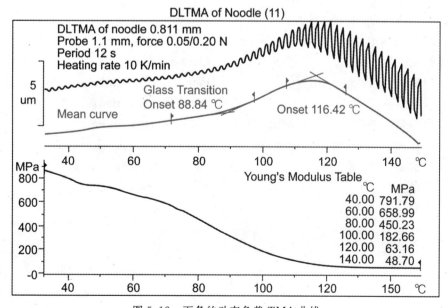

图 5.12 面条的动态负载 TMA 曲线
Figure 5.12 DLTMA curves of the noodle sample

解释 用 1 mm² 平探头,可由动态负载 TMA(DLTMA)曲线计算杨氏模量,因为试样上的压缩应力是已知的。采用球点探头就不行,因为接触面积在测试过程中不断增加。

如图 5.12 所示,直至 110 ℃,面条的厚度是增加的;然后,塑性形变变强。88.8 ℃处的斜率变化可认为是由玻璃化转变引起的,该转变受存在的水量的影响,水在加热时连续解吸,这使得不易进行精确测量。由 DLTMA 曲线的振幅可计算杨氏模量,它在玻璃化转变期间下降最快。

计算 结果列于图中。DLTMA 曲线经空白曲线修正。

结论 在食品加工中,玻璃化转变是一个重要的的性能,可测得的其他信息例如膨胀、塑性行为和模量,不过这些在食品工业领域并不常应用。

Interpretation With the 1 mm² flat probe, the dynamic load TMA (DLTMA) curve allows the calculation of the Young's modulus since the compression stress in the sample is known. With the ballpoint probe, this would not be the case because of the contact area increasing during the course of the measurement.

As shown in figure 5.12, up to approximately 110 ℃ the thickness of the noodle increasesm; afterwards the plastic deformation becomes stronger. The change in slope at 88.8 ℃ is considered to be caused by the glass transition. This transition is influenced by the amount of moisture present, which is continuously being desorbed on heating. This makes a precise measurement difficult. The Young's modulus is calculated from the amplitude of the DLTMA curve and decreases most rapidly during glass transition.

Evaluation The results are listed in the diagram. The DLTMA curve is blank curve corrected.

Conclusion In food technology the glass transition temperature is an important property. The additional information that can be obtained such as expansivity, plastic behavior and modulus is probably unusual in this field of application.

5.12 巧克力的熔融 Melting of Chocolate

样品　牛奶巧克力,瑞士黑巧克力条
Sample　Milk chocolate, Lindt chocolate bar

条件
Conditions

测试仪器:DSC
坩埚:40 μl 铝坩埚,密封。
样品制备:
用锐利的刀将巧克力条切出试样,转移入铝坩埚(勿用手加热!)。试样量 17.33 mg。
测试:
用同一试样进行 DSC 测试。下面括弧内的名称为图 5.13 中所示曲线的名称。
以 10 K/min 由 −30 ℃① 至 60 ℃(原始曲线)
以 20 K/min 由 60 ℃至−30 ℃,恒温 5 min(未显示)

Measuring cell:DSC
Pan:Al 40 μl, hermetically sealed.
Sample preparation:
Sample was cut from a bar with a sharp knife and transferred to an aluminum pan without warming (heat from hands!). Sample weight 17.33 mg.
Measurement:
DSC measurements with the same sample. In brackets are curve names displayed in figure 5.13.
−30 ℃① to 60 ℃ at 10 K/min (original)
60 ℃ to −30 ℃ at 20 K/min, isothermic 5 min (not shown)

以 10 K/min 由 －30 ℃ 至 60 ℃（快速结晶后）	－30 ℃ to 60 ℃ at 10 K/min (after rapid crystallization)
以 1 K/min 由 60 ℃ 至 15 ℃（不显示）	60 ℃ to 15 ℃ at 1 K/min (not shown)
在 15 ℃ 恒温 1 h（不显示）	1 hour isothermic at 15 ℃ (not shown)
以 10 K/min 由 －30 ℃② 至 60 ℃（慢速冷却后）	－30 ℃② to 60 ℃ at 10 K/min (after slow cooling)
注①用约 5 min 从室温降温至 －30 ℃	①Cooling from room temperature to －30 ℃ in approx. 5 min
②用约 4 min 从 15 ℃ 降温至 －30 ℃	②Cooling from 15 ℃ to －30 ℃ in approx. 4 min

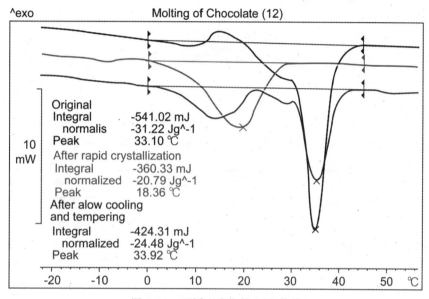

图 5.13 不同巧克力的 DSC 曲线

Figure 5.13 Different DSC curves of the chocolate sample

解释 在本实验中，所用市售牛奶巧克力未做任何进一步制备，以表明多晶型如何随着结晶条件而变化。标识为"Original"的曲线所示，为购买后在 5 ℃ 贮存的市售牛奶巧克力试样的熔融行为。在 10 ℃ 至 20 ℃ 间可清晰辨识出一个放热峰（稳定型的重排）。30 ℃ 以上（口的温度），β-晶型的熔程开始。该晶型在工艺上是通过缓慢的退火获得的。因为升温速率快（如此选择是为了获得较大的峰），没有进一步分辨出各个熔融部分。

在以 20 K/min 将同一试样的熔体快速冷却至 －30 ℃ 后，标识为 "after rapid crystallization" 的曲线呈现新的熔融特征。此时只存在不稳定晶型，同一试样的总熔融热由

Interpretation In this experiment, commercial milk chocolate is used without any further preparation to demonstrate how the crystal modifications vary as a function of the crystallization conditions. The curve labelled 'original' shows the melting behavior of a sample of commercial milk chocolate stored at 5 ℃ after purchase. Between 10 ℃ and 20 ℃ an exothermic transition peak can be clearly recognized (rearrangement to the stable form). Above 30 ℃ (oral temperature) the melting range of the β-modification begins. This modification was achieved technically by gentle annealing. Because of the high heating rate (which was selected to obtain larger peaks), the individual melt fractions are not further resolved.

After faster cooling of the same melt to －30 ℃ at 20 K/min, the curve labelled 'after rapid crystallization' shows new melting characteristics. Only the unstable form is present. The total heat of fusion of the same sample has decreased from 541 mJ (original milk chocolate) to 360 mJ. The rapidly crystallized chocolate no

541mJ（原始牛奶巧克力）下降至 360mJ。快速结晶的巧克力不再满足感官的要求。

longer satisfies the sensory requirements.

现在将同一试样以 1 K/min 的速率由熔体冷却至 15 ℃，恒温 15 min，然后冷却至 −30 ℃。这些退火条件是为了模拟"缓慢而小心的"结晶。该试样的熔融行为由标识为"after slow cooling"的曲线表示。α-晶型已在 5 ℃ 开始熔融，事实上没有进一步的重排（由于升温速率快），而仅有约一半原始 β-晶型存在，在 30 ℃ 之上熔化。

The same sample was now cooled from the melt to 15 ℃ at a rate of 1 K/min, held isothermicly for 15 minutes and then cooled to −30 ℃. These annealing conditions are intended to simulate a 'slow and careful' crystallization. The melting behavior of this sample is shown by the curve labelled 'after slow cooling'. The α-form already starts to melt at 5 ℃. There is virtually no further rearrangement (as a result of the high heating rate). Only about half the original β-modification is now present and melts above 30 ℃.

计算　由峰积分得到估算脂肪晶型相对稳定性所需要的结果：

Evaluation　Peak integration gives the results needed to estimate the relative stability of the fat modification：

	熔融热 Heat of fusion J/g	峰温 Peak temperature ℃
原始巧克力 Original chocolate	31.2	33.1
开始结晶后 After rapid crystallization	20.8	18.4
缓慢结晶后 After slow cooling	24.5	33.9

结论　巧克力原料的质量、工艺参数和组成影响 DSC 熔融曲线的形状。

Conclusion　The raw material quality, the process parameters and the composition of the chocolate have an influence on the shape of the DSC melting curve.

5.13　可可脂的热表征
Thermal Characterization of Cocoa Butter

样品　3 个从瑞士巧克力制造商得到的不同可可脂。
Sample　Three different lots of cocoa butter obtained from Swiss chocolate manufacturers

条件(a)　测试仪器：DSC
Conditions(a)　坩埚：40 μl 铝坩埚，密封。
样品制备：
在 20 ml 烧杯中，在 60 ℃ 使试样熔化达到均匀化。每个试样称重约 10 mg。
测试：
以 1 K/min 由 60 ℃ 降温至 −10 ℃

Measuring cell：DSC
Pan：Al 40 μl, hermetically sealed.
Sample preparation：
The samples are melted at 60 ℃ in 20 ml beakers for homogenization. Approximately 10 mg of each are weighed in.
Measurement：
Cooling from 60 ℃ to −10 ℃ at 1 K/min

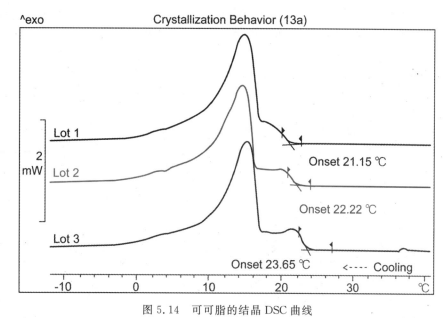

图 5.14　可可脂的结晶 DSC 曲线
Figure 5.14　DSC crystallization curves of cocoa butter

解释　如图 5.14 所示，DSC 曲线的形状表明，有两步结晶，可可脂试样 1、2 和 3 的可再现起始点分别为 21.2 ℃、22.2 ℃和 23.6 ℃。当将生成的晶体在 DSC 仪器中升温时，峰温约为 23 ℃，这表明存在亚稳态产物。已知几毫克物质结晶比克或千克量的温度更低，换言之，对少量试样过冷更大。因此，使用少量试样（在很低温度下结晶）产生最大程度的亚稳产物。

计算　本例只计算了结晶的起始温度。

Interpretation　The shape of the DSC curves indicates a two step crystallization process with reproducible onsets of 21.2 ℃，22.2 ℃ and 23.6 ℃ for the butter samples 1, 2 and 3 respectively. When the crystals formed are heated in the DSC, the peak temperatures are in the region around 23 ℃, which indicates the presence of metastable products. It is known that crystallization of a few milligrams requires lower temperatures than gram or kilogram quantities. In other words, the supercooling is greater with small samples. Therefore the use of small samples (crystallizing at a very low temperature) gives rise to extremely metastable products.

Evaluation　In this case only the onset temperature of crystallization is evaluated.

试样 Sample	试样量 Sample weight mg	起始点 Onset ℃
1	10.82	21.2
2	10.32	22.2
3	10.87	23.6

条件(b、c)　**样品制备**：为了克服若干严重的不符实际的过冷问题，用约 10 g 各在玻璃烧杯中熔化的可可研究其结晶过程。在 60 ℃下

Conditions(b, c)　**Sample preparation**：To overcome the problems of severe unrealistic supercooling, the course of crystallization is studied using approximately 10 g each of melted cocoa butter in glass beakers. After complete fusion at 60 ℃ the beakers are

完全熔化后,将烧杯保持在室温进行结晶,在最初重结晶的 3 天内取样(10.5±0.5g)。

测试:以 1 K/min 由 0 ℃升温至 45 ℃

kept at room temperature for crystallization and samples are taken (10.5 ±0.5g) during the first 3 days of recrystallization.

Measurement:Heating from 0 ℃ to 45 ℃ at 1 K/min

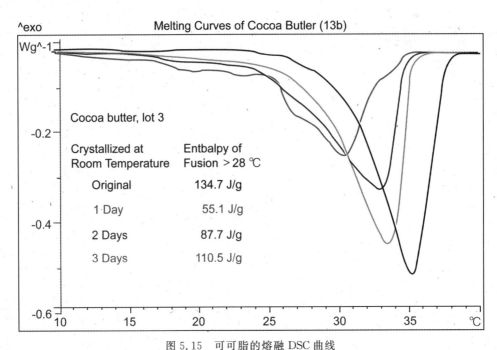

图 5.15　可可脂的熔融 DSC 曲线

Figure 5.15　DSC melting curves of the cocoa butter

解释　图 5.15 所示 DSC 曲线表示试样 3 中晶体形成过程与未加热原始试样的对比。28 ℃以上的积分代表稳定晶型部分的量。熔融热和峰温都朝未熔融原始试样值的方向增大。

Interpretation　The DSC melting curves displayed in figure 5.15 show the course of crystal formation in samples of lot 3 in comparison with the unheated original product. The integral above 28 ℃ represents the fractional amount of the stable modification. Both the heat of fusion and the peak temperature increase towards the values of the unmelted original sample. Both the heat of fusion and the peak temperature increase towards the values of the unmelted original sample.

计算
Evaluation

室温下结晶时间 Ccrystallization time at RT	峰温 Peak temperature ℃	熔融热 Heat of fusion >28 ℃,J/g	$\Delta H/\Delta H_{ori}$ %
1 天 1 day	29.9	55.1	40.9
2 天 2 days	32.5	87.7	65.1
3 天 3 days	32.8	110.5	82.0
原始试样 Original	34.6	134.7	100.0

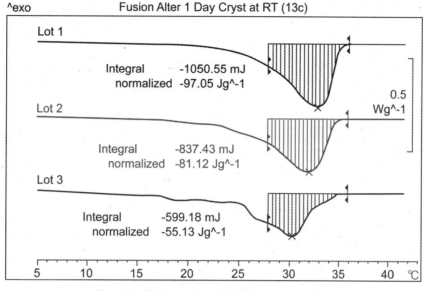

图 5.16 结晶 1 天后的可可脂的熔融 DSC 曲线
Figure 5.16 DSC melting curves of cocoa butter after crystallization 1 day

解释 图 5.16 所示为 3 个不同的可可脂在室温(RT)下于烧杯中结晶一天后测得的熔融 DSC 曲线。每个试样呈现不同的结晶过程。经 1 天结晶后的试样 1 几乎完全为稳定晶型,而试样 3 呈现慢得多的转变。如上所述,28 ℃ 以上的积分代表稳定晶型部分的量。

Interpretation Shown in figure 5.16 are the DSC melting curves of the 3 different lots of cocoa butter measured after they were allowed to crystallize for one day in beakers kept at room temperature (RT). Each lot shows a different course of crystallization. Lot 1 after one day of crystallization is almost completely in the stable modification, whereas lot 3 shows a much slower transformation.

计算
Evaluation

在室温结晶 1 天后 After 1 day of crystallization at RT	峰温 Peak temperature ℃	熔融热 Heat of fusion >28 ℃,J/g	$\Delta H/\Delta H_{ori}$ %
1	32.3	97.0	72.0
2	31.6	81.6	60.2
3	29.9	55.1	40.9
原始试样 3 Original (lot 3)	34.6	134.7	100.0

结论 依赖于结晶条件,在 17 ℃ 至 37 ℃ 之间可观察到熔点(DSC 峰最大值的温度)的变异,这表明没有获得清晰的多晶型。熔融热范围分别从约 50 J/g(可能含有无定形部分)至 140 J/g。
在最佳条件下长期贮存后获得的稳定晶型在 28 ℃ 至 37 ℃ 间熔融。因

Conclusion Depending on the conditions of crystallization, a variety of melting points (the temperature of the DSC peak maximum) between 17 ℃ and 37 ℃ can be observed, which indicates that no distinct polymorph is obtained. The respective heats of fusion range from approximately 50 J/g (possibly containing an amorphous fraction) to 140 J/g.
The stable modification that is obtained after long term storage under optimum conditioning melts between 28 ℃ and 37 ℃. The

此,在该范围测得的熔融热与稳定晶型含量成正比。

heat of fusion measured in this region is therefore proportional to the content of the stable modification.

5.14 熔融行为和氢化作用　Melting Behavior and Hydrogenation

样品 **Sample**	向日葵油 Sunflower oil	
条件 **Conditions**	测试仪器:高压 DSC、DSC	**Measuring cell**:HP DSC,DSC
	坩埚:40 μl 铝坩埚,盖钻孔。	**Pan**:Al 40 μl, with pierced lid.
	样品制备:	**Sample preparation**:
	18.8 mg 向日葵油与约 0.2 mg 用乙醇沾湿的雷尼镍混合,后者起氢化反应催化剂作用。	18.8 mg sunflower oil has been mixed with approx. 0.2 mg Raney nickel moistened with ethanol. This acts later as a catalyst for hydrogenation.
	测试:	**Measurement**:
	在 DSC 中以 5 K/min 由 −50 ℃ 升温至 30 ℃ 以测定原始的熔融行为。在高压 DSC 中氢化,以 5 K/min 由 26 ℃ 升温至 250 ℃。然后在 DSC 仪器中测定最终产物的熔融行为,以 5 K/min 由 25 ℃ 升温至 100 ℃。	Heating from −50 ℃ to 30 ℃ at 5 K/min to determine the original melting behavior in the DSC. Hydrogenation in the HP DSC, heating from 25 ℃ to 250 ℃ at 5 K/min. The melting behavior of the final product then is determined in the DSC, heating from 25 ℃ to 100 ℃ at 5 K/min.
	气氛:	**Atmosphere**:
	DSC 仪器氮气,50 ml/min;高压 DSC 仪器氢气,1 MPa 压力的静止环境。	Nitrogen, 50 ml/min in the DSC; hydrogen, stationary environment at a pressure of 1 MPa in the HP DSC.

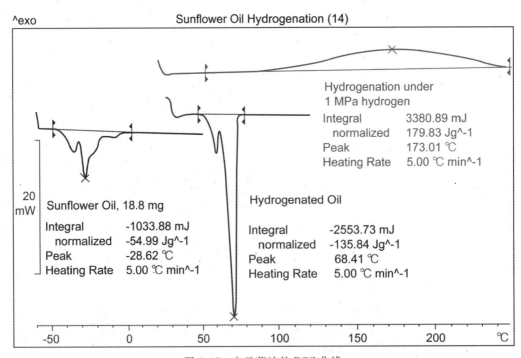

图 5.17　向日葵油的 DSC 曲线

Figure 5.17　DSC curves of the sunflower oil

解释 用 DSC 仪器观察熔融行为来研究氢化效应。随着双键饱和度的增加,熔融峰温由低于 0 ℃升至约 70 ℃,如图 5.17 所示。

Interpretation The effect of hydrogenation is investigated by observing the melting behavior with the DSC. With increasing saturation of the double bonds the temperature of the melting peak increases from below 0 ℃ to almost 70 ℃, as shown in figure 5.17.

计算 放热氢化反应发生在 40 ℃至 250 ℃间,反应热为 180 J/g

Evaluation The exothermic hydrogenation takes place between 40 ℃ and 250 ℃ with a heat of reaction of 180 J/g.

试样 Sample	熔融热 Heat of fusion J/g	峰温 Peak temperature ℃
原始样 Native	55.0	−28.6
氢化后 Hydrogenated	135.8	68.4

结论 熔融行为是食用油、脂肪最重要的特性之一,影响其味觉性能。DSC 不仅可测定因氢化作用导致熔融范围向较高温度移动,而且化学反应本身就能在测试池内进行。

Conclusion The melting behavior is one of the most important characteristics of edible oils and fats and affects organoleptic properties. DSC not only shows the shift of the melting range to higher temperature as a result of hydrogenation, but also allows the chemical reaction itself to be performed in the measuring cell.

5.15 植物油的结晶 Crystallization of Vegetable Oils

样品 菜籽油、大豆油(Homa 牌)、橄榄油(Dante 牌)、棕榈油
Sample Rape seed oil, Soybean oil (Homa brand), Olive oil (Dante brand), Palmoil

条件
测试仪器:DSC
坩埚:40 μl 铝坩埚,密封。
样品制备:未作特别制备。
测试:
以 10 K/min 由 50 ℃降温至−100 ℃

Conditions
Measuring cell:DSC
Pan:Al 40 μl, hermetically sealed.
Sample preparation:No special sample preparation.
Measurement:
Cooling from 50 ℃ to −100 ℃ at 10 K/min

解释 如图 5.18 所示,橄榄油在低于−10 ℃时结晶。含有部分饱和脂肪酸的甘油三酯部分在−10 ℃至−35 ℃间结晶。橄榄油的主要部分即含 3 个油酸单元的甘油三酯(70%)在更低的温度结晶,这可由−45 ℃的结晶峰辨认(稳定 α-晶型的文献温度为−35 ℃)。

市售棕榈油在低于+15 ℃时结晶。它含有高百分比的饱和脂肪酸

Interpretation As shown in figure 5.18, Olive oil crystallizes below −10 ℃. The triglyceride fraction, with in part saturated fatty acids, crystallizes between −10 ℃ and −35 ℃. The main fraction of olive oil, the triglyceride with 3 oleic acid units (70%), crystallizes at lower temperatures and is recognizable as a crystallization peak at −45 ℃ (a temperature of −35 ℃ is given in the literature for the stable α-form).

Commercial palm oil crystallizes below +15 ℃. It has a high percentage of saturated fatty acids (50% C12, 18% C14) and

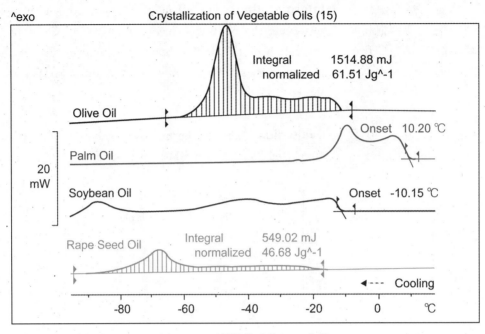

图 5.18 不同植物油的 DSC 曲线
Figure 5.18 DSC curves of different vegetable oils

(50% C12、18% C14),因而熔点高;8%不饱和油酸使熔点降低到室温以下。

大豆油含较多的饱和脂肪酸(10% C16),还有较高百分比的一元和二元不饱和脂肪酸(具有营养价值)。用 HPLC 法分析甘油三酯试样表明,主要为含饱和脂肪酸的部分,有一些不饱和脂肪酸的部分,还有一些高度不饱和脂肪酸部分。由 DSC 曲线不同的结晶温度可观察到这 3 部分甘油三酯。大豆油中高比例的亚麻酸(50%)产生高比例的含三个亚麻酸单元的甘油三酯(熔点－45 ℃),呈现－42 ℃的峰。

菜籽油实际上仅含有不饱和脂肪酸(68% C18 一元酸、25% C18 二元酸、10% C18 三元酸),在很低的温度结晶。只有 5%的脂肪酸是饱和的,呈现为－20 ℃至－40 ℃间的"小峰"。即使无法清晰地鉴别脂肪酸或甘油三酯部分,DSC 还是可以快速表征油/脂肪的。

hence a high melting point; 8% unsaturated oleic acid lowers the melting point to values below room temperature.

Soybean oil contains a significant fraction of saturated fatty acids (10% C16) and also a high percentage of monobasic and dibasic unsaturated fatty acids (nutritionally valuable). Triglyceride specimens analyzed by HPLC show fractions with predominantly saturated fatty acids, with some unsaturated fatty acids and also with highly unsaturated fatty acids. These 3 triglyceride fractions are visible in the DSC curve at different crystallization temperatures. The high fraction of linolenic acid in soybean oil (50%) leads to a high fraction of glycerides with three linolenic acid units (melting point -45 ℃), which is shown as a peak at -42 ℃.

Rape seed oil contains virtually only unsaturated fatty acids (60% C18 monobasic, 25% C18 dibasic, 10% C18 tribasic) and crystallizes at a very low temperature. Only 5% of the fatty acids are saturated and this is indicated by 'small peaks' between -20 ℃ and -40 ℃. Even though DSC cannot clearly identify fatty acids or triglyceride fractions, rapid characterization of the oils/fats is possible.

计算 Evaluation	样品 Sample	试样量 Sample weight mg	结晶热 Heat of crystallization J/g	结晶起始点 Onset of crystallization ℃
	橄榄油 Olive oil	24.63	61.5	−10.3
	棕榈油 Palm oil	7.23	91.3	+10.2
	大豆油 Soybean oil	27.67	30.9	−10.2
	菜籽油 Rape seed oil	11.76	46.7	−17.1

结论 DSC 是研究食用油结晶行为的快速手段。为了对比,应取大致相同的试样量,例如(20±5) mg。试样量可能影响过冷:试样量越少,过冷程度越大。

Conclusion DSC is a rapid means of investigating the crystallization behavior of edible oils. For comparison purposes approximately identical sample sizes should be taken, e.g. (20±5) mg. The sample size may influence the supercooling: the smaller the sample size the greater the degree of supercooling.

5.16 棕榈油的液相含量和滴点
Liquid Fraction and Dropping Point of Palm Oils

样品 棕榈油和棕榈油馏分(如棕榈油 olein 或棕榈油 strearin)
Sample Palm oils and fractionated palm oils such as palm olein or palm strearin

条件
Conditions

测试仪器:DSC
Measuring cell:DSC

坩埚:40 μl 铝坩埚,密封。
Pan:Al 40 μl, hermetically sealed.

样品制备:
Sample preparation:

为了更均匀,在进样前用刮板搅拌脂肪。
For better homogeneity the fat was stirred with a spatula before sampling.

测试:
Measurement:

以 10 K/min 由−50 ℃升温至 80 ℃(约用 10 min 由室温降温至−40 ℃)
Heating from −40 ℃ to 80 ℃ at 10 K/min (cooling from RT to −40 ℃ in approximately 10 min).

气氛:氮气,100 ml/min
Atmosphere:Nitrogen,100 ml/min

解释 由熔融峰的部分积分计算得到液相含量(转化率曲线),如图 5.19 右边的液相含量曲线所示。获得正确结果的前提是:用 J/g 表示的所有含量的熔融热是相同的,在 DSC 开始测量时结晶是完全的(无非晶相残留)。如果有的试样的熔融热较低,则认为存在起始的非晶相。要计算这种经修正的液相含量,输入最大值作为"文献值"(这里

Interpretation The liquid fraction is calculated as the partial integral of the fusion peak (conversion curve), as shown by the liquid fraction curves in the right diagram of figure 5.19. Prerequisites for correct results are: the heat of fusion in J/g of all fractions is the same and that the crystallization at the DSC starting temperature is complete (no amorphous phase remaining). If some samples exhibit a lower heat of fusion, it is assumed that there is an initial amorphous phase. To calculate such corrected liquid fractions, the highest value is entered as a 'literature value' (here −85 J/g for palm stearine) and ther curve begins

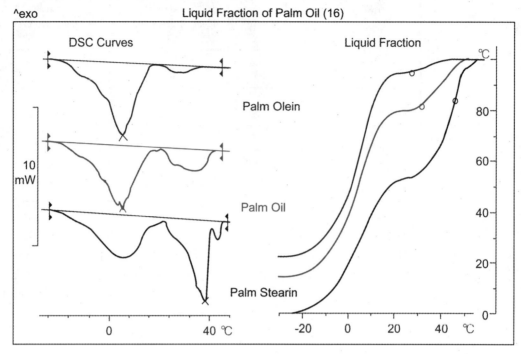

图 5.19 不同棕榈油的 DSC 曲线和液相含量曲线
Figure 5.19 DSC curves and liquid fraction curves of different palm oils

对棕榈油 stearine 为 -85 J/g），曲线从大于零开始。

above zero.

液相含量曲线上的圆圈为用 FP83 测试仪器测得的滴点温度。

The circles in the liquid fraction curves show the dropping point temperatures obtained with the FP83 measurement module.

计算
Evaluation

样品 Sample	试样量 Sample weight mg	25 ℃时液相含量 Liquid fraction at 25 ℃ %	最后峰温度 Temperature of last peak ℃
棕榈油 olein Palm olein	19.03	94.9	37.7
棕榈油 Palm oil	19.82	79.8	40.7
棕榈油 stearin Palm stearin	21.14	53.1	53.7

条件 测试仪器：FP83 型仪器，放置于低温冷冻室，使仪器以 -10 ℃ 为起始温度。

样品制备：用刮板将石蜡状粘稠的样品压入滴点杯（直径 2.8 mm 敞口杯）中。

测试：以 1 K/min 由 -10 ℃升温至 60 ℃（约用 20 min 由室温降温至 -10 ℃）

Conditions Measuring cell：FP83, placed in a deep freezer to allow a starting temperature of -10 ℃

Sample preparation: Samples of waxy consistency were pressed into the dropping point cup (2.8 mm of diameter, opening) with a spatula.

Measurement: Heating from -10 ℃ to 60 ℃ at 1 K/min (cooling from RT to -10 ℃ in approx. 20 minutes)

| 气氛:空气,静止环境,无流动 | Atmosphere: Air, stationary environment, no flow rate |

解释 缓慢加热时流出标准滴点试样杯第1滴物质时的温度定义为滴点。也在图 5.19 中的液相含量曲线上用圆圈标注了测得的滴点,以兹比较。

Interpretation The dropping point is defined as the temperature at which the first drop of substance flows out of the opening of a standard dropping point sample cup on slow heating. The measured dropping points are also noted with circles in the liquid fraction curves in figure 5.19 for comparison purposes.

计算
Evaluation

样品 Sample	滴点 Dropping point ℃	标准偏差 Standard deviation ℃
棕榈油 olein Palm olein	27.3	0.09
棕榈油 Palm oil	31.2	0.15
棕榈油 stearin Palm stearin	45.6	0.17

注:滴点为4次测定的平均值。
Note: The dropping points were the mean value of 4 determinations.

结论 不同来源的棕榈油和不同馏分的棕榈油例如棕榈油 olein 或棕榈油 stearin 有不同的熔融和结晶行为。因此,熔融和结晶 DSC 曲线能表征这些食用脂肪。滴点相同的脂肪可能有不同的液相含量,可用 DSC 来区别。

Conclusion Palm oils of different origin and fractionated palm oils such as palm olein or palm stearin have different melting and crystallization behavior. The DSC melting and crystallization curves therefore allow the characterization of such edible fats. Fats with identical dropping points can have different liquid fraction data, i.e. they can be distinguished by DSC.

5.17 植物脂肪的氧化 Oxidation of Vegetable Fats

样品 Homa 牌大豆油(每 100 ml 油含 28 g 维他命 E)
棕榈油脂肪
Sample Soybean oil, Homa brand (28 g vitamin E per 100 ml oil)
Palm fat

条件 测试仪器:DSC
Conditions 坩埚:40 μl 铝坩埚,无盖。
样品制备:
无专门的试样制备,但坩埚无盖!
测试:
以 15 K/min 由 50 ℃升温至 390 ℃
气氛:氧气或氮气,100 ml/min

Measuring cell: DSC
Pan: Al 40 μl, no lid.
Sample preparation:
No special sample preparation, but pan without lid!
Measurement:
Heating from 50 ℃ to 390 ℃ at 15 K/min
Atmosphere: Oxygen or nitrogen, 100 ml/min

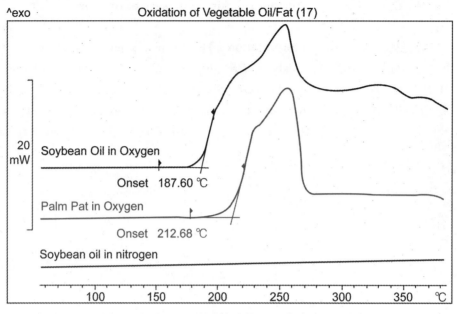

图 5.20 植物脂肪的 DSC 曲线
Figure 5.20　DSC curves of the vegetable fats

解释　食用油和脂肪的自动氧化对含脂肪产品的贮存和加工有负面影响。例如在烘烤和煎炸时，150 ℃以上就开始氧化，如图 5.20 所示。大豆油在在氮气中不呈现任何可观察到的分解信号。

Interpretation　The autooxidation of edible oils and fats has a negative influence on the storage and processing of products which contain fat. Above 150 ℃, for instance in roasting or frying, oxidation sets in, as shown in figure 5.20. Soybean oil under nitrogen shows no visible signs of decomposition.

计算
Evaluation

样品 Sample	试样量 Sample weight mg	气氛 Atmosphere	氧化开始 Start of oxidation ℃
大豆油 Soybean oil	1.13	O_2	188
大豆油 Soybean oil	2.06	N_2	—
棕榈油 Palm fat	0.57	O_2	213

结论　氧在足够高温度时侵蚀有机物质。DSC 起始温度越高，稳定性就越好。可相互区别使用过的脂肪与新鲜脂肪。

Conclusion　Oxygen attacks organic substances at sufficienly high temperatures. The higher the DSC onset temperature the better the stability. Used fats and fresh fats can be distinguished from each other.

5.18　乙醇/水混合物　Ethanol/Water Mixtures

样品　99.5％乙醇，用蒸馏水制备 60％、40％和 20％体积浓度的乙醇溶液；
爱尔兰都柏林 Tullamore Dew 公司 Tullamore Dew 牌威士忌，40％体积浓度

Sample		Ethanol 99.5%, distilled water to prepare 60%, 40% and 20% ethanol solutions by volume;	
		Whiskey, Tullamore Dew brand, Tullamore Dew Company, Dublin, Ireland, 40% volume	
条件	测试仪器：DSC	**Measuring cell**：DSC	
Conditions	坩埚：40 μl 铝坩埚，密封。	**Pan**：Al 40 μl, hermetically sealed.	
	样品制备：	**Sample preparation**：	
	在室温称量试样（16±2）mg、密封，将坩埚放入预先冷却的 DSC 样品皿中，以最大速率冷冻，并在 −160 ℃恒温 5～10 min	Samples weighed at room temperature (16±2) mg, hermetically sealed, frozen at max. cooling rate by inserting the pan into the precooled DSC cell and kept isothermicly for 5−10 minutes at −160 ℃.	
	测试：	**Measurement**：	
	以 10 K/min 由 −160 ℃升温至 +20 ℃	Heating from −160 ℃ to +20 ℃ at 10 K/min	

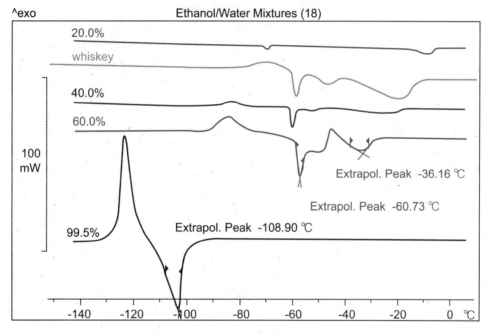

图 5.21 乙醇/水混合物的 DSC 曲线

Figure 5.21 DSC curves of the ethanol/water mixtures

解释 迄今已由激光拉曼或质子磁共振光谱测得的乙醇/水混合物的热力学"异常"，也可用 DSC 观察到。

99.5%乙醇试样在 −123 ℃呈现放热结晶峰，在 −105 ℃呈现熔融峰。该曲线形状只有在 98%至 100%浓度范围可观察到。

60%乙醇/水混合物在 −85 ℃呈现放热峰。对 10%至 85%的乙醇/水混合物可观察到类似的行为。如同 99.5%乙醇试样的情况，认为是乙

Interpretation The thermodynamic 'abnormalities' of ethanol/water mixtures, which have been measured until now by laser raman or proton magnetic resonance spectroscopy, can also be observed by DSC.

The 99.5% ethanol sample shows an exothermic crystallization peak at −123 ℃ and a melting peak at −105 ℃. This curve shape is observed only at concentrations between 98% and 100%.

The 60% ethanol/water mixture shows an exothermic peak at −85 ℃. Similar behavior is observed for ethanol/water mixtures from 10% to 85%. As in the case of the 99.5% ethanol sample, recrystallization of the ethanol/water mixture is assumed. The

醇/水混合物的重结晶。混合物的熔融峰在-60.7 ℃。最大值在-48 ℃处的吸热过程（看似肩形，立即跟随着熔融峰），与水-乙醇很大的相互作用有关。60%乙醇溶液体积最小，自由能最大，导致大量形成水/乙醇键，其中一些在此温度破裂。

乙醇浓度53%至75%，在-45 ℃出现一个放热峰。在该温度，熔融开始前发生水的重结晶，水分的熔点为-38 ℃。乙醇浓度较低时，熔点移至较高温度。

40%的溶液呈现所有的峰，包括水的重结晶，不过较弱。

20%的乙醇溶液仅呈现两个纯组分的熔点。没有发生重结晶。

可利用熔融图，或更准确的是用-50 ℃处的吸热峰，来测定亦已贮存很久的饮品（威士忌）的年限，该研究工作已公开发表。

结论 本例是为了说明，一方面，可用DSC进行监测复杂的冷冻和熔化过程，另一方面，所有的相转变是显而易见的。最重要的是，用DSC可研究游离水分和水产品的冷冻和熔化行为。

melting peak of the mixture lies at -60.7 ℃. An endothermic process with a maximum at -48 ℃, observed as a shoulder immediately following the melting peak, involves the destruction of the powerful water-ethanol interactions. A 60% ethanol solution has a minimum volume and a maximum free energy which leads to extensive formation of water/ethanol bonds, some of which are broken at this temperature.

An exothermic peak at -45 ℃ appears at ethanol concentrations of 53% to 75%. At this temperature, a recrystallization of the water occurs before the melting begins. The melting point of the water fraction is -38 ℃. At lower ethanol concentrations, it is shifted to higher temperatures.

The 40% solution shows all the peaks including that of the recrystallization of water, which however is weaker.

The 20% ethanol solution shows only the melting peaks of the two pure components. Recrystallizations do not occur.

Work has been published which makes use of the melting diagram or to be more accurate, the endothermic peak at -50 ℃, to determine the age of beverages (whiskey) that have been stored for a long time.

Conclusion This example is intended to demonstrate that on the one hand even complex freezing and thawing processes can be monitored by DSC, and that on the other hand, all the phase transitions are readily apparent. Above all, the freezing and thawing behavior of free water and aqueous products can be investigated by DSC.

5.19 塑料薄膜的鉴别　Identification of Plastic Films

样品 高密度聚乙烯(PE-HD)/聚酰胺6(PA6)和低密度聚乙烯(PE-HD)/聚酰胺6(PA6)复合薄膜
Sample PE-HD/PA6 and PE-LD/PA6 composite films

条件 测试仪器：DSC
Conditions 坩埚：40 μl铝坩埚，盖钻孔。
样品制备：
由薄膜冲出一个圆片，平放入坩埚。

测试：
以10 K/min由30 ℃升温至290 ℃
气氛：氮气，50 ml/min

Measuring cell：DSC
Pan：Al 40 μl, lid pierced.
Sample preparation：
A disk is punched out of the film and placed flat in the pan.

Measurement：
Heating from 30 ℃ to 290 ℃ at 10 K/min
Atmosphere：Nitrogen, 50 ml/min

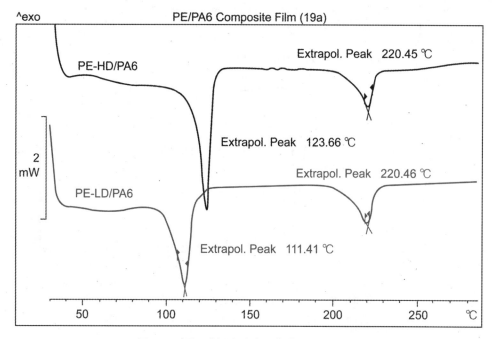

图 5.22 聚乙烯/聚酰胺 6 薄膜的 DSC 曲线
Figure 5.22　DSC curves of the PE/PA6 films

解释　图 5.22 中上面的曲线是从 PE-HD(高密度聚乙烯)和 PA6(聚酰胺 6)复合物测得的,该复合物通常作最终产品的包装材料(食品袋)。PE-HD 在约 125 ℃ 熔融,PA6 在约 220 ℃ 熔融。

在下面的曲线上,PE 的熔融峰要低 15 ℃。该薄膜也是 PA/PE 复合物,但用的是在 110 ℃ 熔融的 PE-LD(低密度聚乙烯)。该薄膜通常用作腊肠和奶酪的包装材料。

另一种适用的薄膜,由聚丙烯和聚酯(PET)制造,还适用于微波炉,可容易地将薄膜区分,因为 PET 的熔点(DSC 曲线峰温)是 256 ℃(玻璃化转变在 69 ℃),聚丙烯的熔点是 165 ℃。

Interpretation　The upper curve in figure 5.22 is obtained from a composite of PE-HD (PolyEthylene of High Density) and PA6 (polyamide 6) that is normally used for the packaging of finished products (cooking bags). PE-HD melts at approximately 125 ℃ and PA6 at 220 ℃.

In the bottom curve the PE melting peak is 15 ℃ lower. This film is also a PA/PE composite, but uses PE-LD (PolyEthylene of Low Density) which melts at 110 ℃. The film is a common packaging material for sausages and cheese.

Another suitable film, made of polypropylene and polyester (PET), which is also usable in microwave ovens, could easily be distinguished from the present film since the melting 'point' (DSC peak temperature) of PET is 256 ℃ (with a glass transition at 69 ℃) and that of polypropylene 165 ℃.

计算
Evaluation

样品 Sample	试样量 Sample weight mg	峰温 Peak temperature ℃	
		PE	PA
PE-HD/PA6	2.38	123.7	220.4
PE-LD/PA6	2.64	111.4	220.5

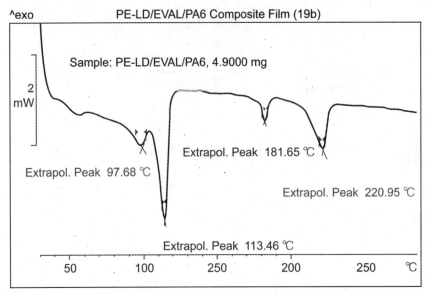

图 5.23 低密度聚乙烯/乙烯-乙烯醇共聚物/聚酰胺 6 薄膜的 DSC 曲线
Figure 5.23 DSC curves of the PE-LD/EVAL/PA6 film

解释 图 5.23 为低密度聚乙烯/乙烯-乙烯醇共聚物/聚酰胺 6 薄膜的 DSC 曲线。该薄膜为三层复合物，在 PE-LD 和 PA6 之间有一层 EVAL(乙烯-乙烯醇共聚物)不透气层。EVAL 的熔点约为 180 ℃。随着乙烯含量的增大（由 32% 至 44%），熔点下降。含量根据需要而改变。不过，不断增大的不透气性能伴随着加工性能的不断下降。
在 98 ℃还有一个添加剂的熔融峰。这种薄膜用于对隔氧要求高的奶酪和半成品食品行业（真空包装或含保护性气体）。

Interpretation Figure 5.23 shows the DSC curves of PE-LD/EVAL/PA6 film. This film is a three-layer composite which has a gas-impermeable layer of EVAL (Ethylene-Vinyl ALcohol copolymer) between the PE-LD and the PA6. The melting point of EVAL is approximately 180 ℃. With increasing ethylene content (32% to 44%), the melting point decreases. The content varies depending on the requirements. However, increasing barrier capability is accompanied by poorer processability.

There is also a melting peak of an adhesive at 98 ℃.
Such films are used in the food sector for cheese and semi-finished products with high barrier requirements towards oxygen (evacuated packages or those with a protective atmosphere).

计算
Evaluation

组分 Component	添加剂 Adhesive	PE-LD	EVAL	PA6
峰温 Peak temperature ℃	97.7	113.5	181.6	221.0

结论 塑料材料的鉴别对于竞争性产品的检验以及成品的质量控制是必不可少的。DSC 不仅可鉴别聚合物的主要组分，还可检测次要成分例如添加剂，如果它们具有不同于聚合物的熔融行为。

Conclusion The identification of plastic materials is essential for the examination of competitive products as well as for the quality control of finished goods. DSC can not only identify the major components of polymers, but can also detect minor components such as adhesives, if these have a melting behavior which differs from that of the polymers.